**INTRODUCTION TO
EFFICIENT ELECTRICAL
SYSTEMS DESIGN**

INTRODUCTION TO

EFFICIENT

ELECTRICAL

SYSTEMS

DESIGN

By
Stephen Ayraud, P.E.
and
Albert Thumann, P.E., C.E.M.

THE FAIRMONT PRESS, INC.
P.O. Box 14227, Atlanta, Georgia 30324

PRENTICE-HALL, INC.
Englewood Cliffs, New Jersey 07632

Introduction to Efficient Electrical Systems Design

©Copyright 1985 by The Fairmont Press, Inc.
All rights reserved. No part of this publication may be reproduced, stored in a retrieval system, or transmitted, in any form or by any means, electronic, mechanical, photocopying, recording, or otherwise, without the prior written permission of the publisher.

This edition published 1985 by Prentice-Hall, Englewood Cliffs, NJ 07632

ISBN 0-13-481318-9

Printed in the United States of America

10 9 8 7 6 5 4 3 2 1

Library of Congress Cataloging in Publication Data

Ayraud, Stephen, 1954-
    Introduction to efficient electrical systems design.

    Includes index.
    1. Factories—Electric equipment. 2. Buildings—
Electric equipment. 3. Electric engineering.
I. Thumann, Albert, II. Title.
TK4035.F3A97   1985   621.3   83-49500
ISBN 0-915586-98-3

ISBN 0-13-481318-9   01

Prentice-Hall International, Inc., *London*
Prentice-Hall of Australia Pty. Limited, *Sydney*
Editora Prentice-Hall do Brasil, Ltda., *Rio de Janeiro*
Prentice-Hall Canada Inc., *Toronto*
Prentice-Hall Hispanoamericana, S.A., *Mexico*
Prentice-Hall of India Private Limited, *New Delhi*
Prentice-Hall of Japan, Inc., *Tokyo*
Prentice-Hall of Southeast Asia Pte. Ltd., *Singapore*
Whitehall Books Limited, *Wellington, New Zealand*

This book is dedicated to
my father
who showed me how to install
electrical systems.

Albert Thumann

FOREWORD

The understanding of electrical system design has become increasingly important, not only to the electrical designer, but to safety, plant and project engineers as well. With the advent of high energy costs, plant and project engineers have needed to become more aware of electrical systems. Both safety and energy efficiency will be covered in this text along with practical application problems for industrial and commercial electrical design.

CONTENTS

Chapter 1

ELECTRICAL BASICS

The field of electrical engineering is a large and diverse one. Often included under the general title of electrical engineer are the fields of: electronics, semiconductors, computer science, power, lighting and electro-magnetics. The focus of this book is on the consulting or plant electrical engineer whose responsibilities include facility power distribution and lighting.

The following chapter provides a brief review of basic concepts which serve as background for the electrical engineer. A thorough understanding of these concepts, while helpful, is not essential to understanding the remainder of this book.

ELECTRICAL UNITS

Table 1-1 and the following text provide definitions of the basic electrical quantities.

Table 1-1. Electrical Quantities in MKS Units

Quantity	Symbol	Definition	Unit
Force	f	push or pull	Newton
Energy	w	ability to do work	joule or watt-second
Power	p	energy/unit of time	watt
Charge	q	integral of current	coulomb
Current	i	rate of flow of charge	ampere
Voltage	v	energy/unit charge	volt
Electric field strength	E	force/unit charge	volt/meter
Magnetic flux density	B	force/unit charge momentum	tesla
Magnetic flux	ϕ	integral of magnetic weber flux density	weber

1

"Force" A force of 1 newton is required to cause a mass of 1 kilogram to change its velocity at a rate of 1 meter per second per second.

"Energy" Energy in a system is measured by the amount of work which the system is capable of doing. The joule or watt-second is the energy associated with an electromotive force of 1 volt and the passage of one coulomb of electricity.

"Power" Power measures the rate at which energy is transferred or transformed. The transformation of 1 joule of energy in 1 second represents an average power of 1 watt.

"Charge" Charge is a "quantity" of electricity. The coulomb is defined as the charge on 6.24×10^{18} electrons, or as the charge experiencing a force of 1 newton in an electric field of one volt per meter, or as the charge transferred in 1 second by a current of 1 ampere.

"Current" The current through an area is defined by the electric charge passing through per unit of time. The current is the net rate of flow of positive charges. In a current of 1 ampere, charge is being transferred at the rate of 1 coulomb per second.

"Voltage" The energy-transfer capability of a flow of electric charge is determined by the potential difference or voltage through which the charge moves. A charge of 1 coulomb receives or delivers an energy of 1 joule in moving through a voltage of 1 volt.

"Electric Field Strength" Around a charge, a region of influence exists called an "electric field." The electric field strength is defined by the magnitude and direction of the force on a unit positive charge in the field (i.e., force/unit charge).

"Magnetic Flux Density" Around a moving charge or current exists a region of influence called a "magnetic field." The intensity of the magnetic effect is determined by the magnetic flux density which is defined by the magnitude and direction of a force exerted on a charge moving in the field with a certain velocity. A force of 1 newton is experienced by a charge of 1 coulomb moving with a velocity of 1 meter per second normal to a magnetic flux density of 1 tesla.

"Magnetic Flux"   Magnetic flux quantity, in webers, is obtained by integrating magnetic flux density over an area.

RESISTANCE

If a battery is connected with a wire to make a complete circuit, a current will flow. (See the schematic representation in Figure 1-1.) The current that flows is observed to be proportional to the applied voltage. The constant that relates the voltage and current is called "resistance." If the symbol v represents volts, the symbol i represents current and the symbol R represents resistance (measured in ohms-Ω) the relationship can be expressed by the equation:

FORMULA 1-1 $v = Ri$

This expression is called Ohm's Law.

Since voltage is the energy per unit charge and current is the charge per unit time, the basic expression for electrical energy per unit time or power is:

FORMULA 1-2 $P = vi = i^2R$

Consequently, resistance is also defined as a measure of the ability of a device to dissipate (in the form of heat) power.

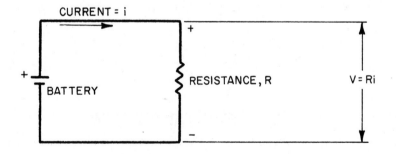

FIGURE 1-1 **OHM'S LAW REPRESENTATION**

CAPACITANCE

Now let's connect the battery to two flat plates separated by a small air space between them. (See the schematic representation in Figure 1-2.) When a voltage is applied, it is observed that a positive charge appears on the plate connected to the positive terminal of the battery and a negative charge appears on the plate connected to the negative terminal. If the battery is disconnected, the charge persists. Such a device which stores charge is called a capacitor.

If a device called a signal generator, which generates an alternating voltage, is installed in place of the battery, the current is observed to be proportional to the rate of change of voltage. The relationship can be expressed by the equation:

FORMULA 1-3 $i = c \, dv/dt$

where C is a constant called "capacitance" (measured in farads) and dv/dt is differential notation representing the rate of change of voltage.

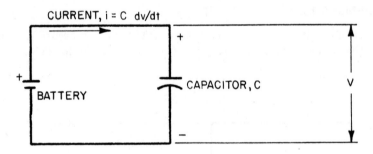

FIGURE 1-2 CAPACITANCE LAW

INDUCTANCE

If the signal generator is placed in a circuit in which a coil of wire is present, it is observed that only a small voltage is required to maintain a steady current. (See the schematic representation in Figure 1-3.) However, to produce a rapidly changing current, a relatively large voltage is required. The voltage is observed to be proportional to the rate of change of the current and can be expressed by the equation:

FORMULA 1-4 $v = L\, di/dt$

where L is a constant called "inductance" (measured in henrys— H) and di/dt is differential notation representing the rate of change of current.

Additionally, when a direct current is removed from an inductor the resulting magnetic field collapses thereby "inducing" a current in an attempt to maintain the current flow. Consequently, inductance is a measure of the ability of a device to store energy in the form of a magnetic field.

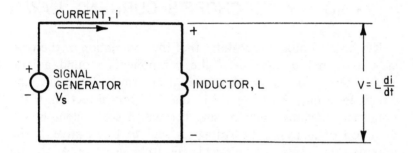

FIGURE 1-3   INDUCTANCE LAW

CIRCUIT LAWS

To ease the analysis of complex circuits, circuit laws are utilized. These circuit laws allow voltages and currents to be calculated if only some of the circuit information is known. The two most famous circuit laws are Kirchoff's Current and Voltage Laws. Kirchoff's Current Law states that the sum of the currents flowing into a common point (or node) at any instant is equal to the sum of the currents flowing out.

If current flowing into a node is taken as positive and current flowing out of a node is taken as negative, the summation of all the currents at a node is zero. If the circuit represented by

Figure 1-4 is analyzed, Krichoff's Law would be utilized as follows:

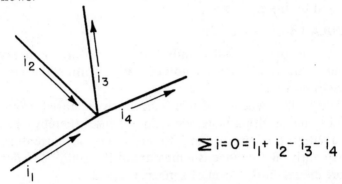

$$\sum i = 0 = i_1 + i_2 - i_3 - i_4$$

FIGURE I-4 **KIRCHOFF'S CURRENT LAW**

Kirchoff's Voltage Law states that the summation of the voltages measured across all of the components around a loop equals zero. To analyze a circuit using the voltage law, the circuit loop must be traversed in one arbitrary direction. If the potential increases when passing through a component in the direction of analysis, the voltage is said to be positive. If the potential decreases, the voltage is said to be negative. Analyzing the circuit of Figure 1-5 gives the following:

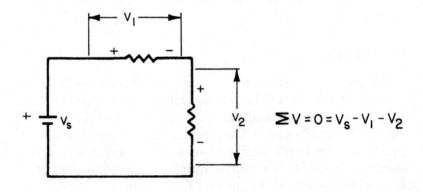

$$\sum V = 0 = V_s - V_1 - V_2$$

FIGURE I-5 **KIRCHOFF'S VOLTAGE LAW**

ALTERNATING CURRENT

Although direct current finds uses in some semiconductor circuitry, the primary focus of electrical design is concerned with alternating currents. Among the many advantages of utilizing alternating currents are: the voltage is easily transformed up or down, an alternating current varying at a prescribed frequency provides a dependable time standard for clocks, motors, etc. and alternating patterns (i.e., sound and light waves) occur in nature and consequently provide the basis for analysis of signal transmission.

An alternating current, or sinusoid, can be represented by the equation (see Figure 1-6):

FORMULA 1-5 $a = A \cos (\omega t + \alpha)$

where a = instantaneous value
      A = amplitude or maximum value
      ω = frequency in radians per second (omega)
      t = time in seconds
      α = phase angle in radians (alpha)

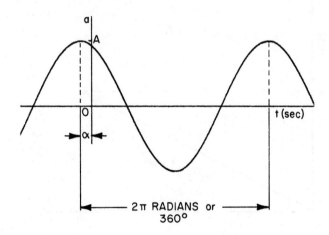

Figure 1-6. Sinusoid Representation

Note that frequency ω is related to frequency (f) in cycles per second or Hertz (Hz) by:

FORMULA 1-6 $f = \omega/2\pi$

(In the United States, the common power distribution frequency is 60 Hz.)

The phase angle α represents the difference between the reference time t = 0 and the time that the peak amplitude A occurs.

It is convenient to represent Formula 1-5 and Figure 1-6 by a phasor diagram as shown in Figure 1-7.* In this figure, the sinusoid is represented as a complex vector rotating with a frequency ω from an initial phase angle α. Such a concept allows the sinusoid to be represented by complex constants (composed of real and imaginary parts) instead of functions of time and also allows phasors to be added together using the rules of complex algebra.

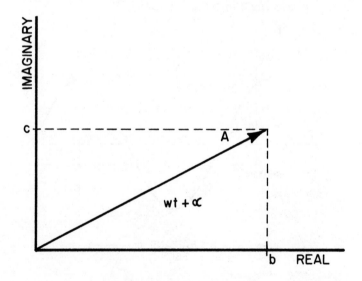

Figure 1-7. Phasor Representation of a Sinusoid

Using phasor notation for the sinusoid represented by Figure 1-7, we get:

FORMULA 1-7 $a = b + jc = Ac(j\alpha) = A\angle\alpha$

Consequently, the term $A\angle\alpha$ is a simplified representation of Formula 1-5 for a given frequency.

*For more information on the phasor concept, see *Circuits, Devices and Systems*, R.J. Smith, John Wiley and Sons, Inc., 1971

IMPEDANCE

To represent the effect that capactive and inductive elements have on the current in a circuit, the concept of impedance is introduced.

In a circuit with capacitive and inductive elements present, Ohm's law is modified to be the following, where all symbols are represented by phasors:

FORMULA 1-8 $Z = V/I$

where Z is measured in Ohms.
In a series circuit, Z is defined as:

FORMULA 1-9 $Z = \sqrt{R^2 + X^2}\angle\Theta$

where R = resistance
X = reactance
Θ = the angle that the voltage phasor leads or lags the current phasor

REACTANCE

An inductive element opposes a change in alternating current. We say that the voltage across an inductive element leads by 90° the current through it. Its reactance is given by:

FORMULA 1-10 $X_L = V_L/I_L = \omega L\angle 90° = 2\pi f L\angle 90°$

The voltage across a capacitive element lags by 90° the current through it. Its reactance is given by:

FORMULA 1-11 $X_C = V_C/I_C = 1/\omega C\angle{-90°} = 1/2\pi f C\angle{-90°}$

Note that the voltage and current are in phase across a purely resistive element.

POWER

In a circuit containing resistive elements only, the voltage and current are in phase and power is calculated by Formula 1-2. When inductive and/or capacitive elements are present, however, the energy is not dissipated in these elements but is stored and returned to the circuit every half cycle. The current and voltage are not in phase in reactive elements and consequently Formula 1-12 must be used to calculate power consumption.

FORMULA 1-12 $P = VI \cos \Theta$ watts

The difference between the power in watts and the "volt-amperes" is the product of a quantity termed the power factor which is calculated by Formula 1-13.

FORMULA 1-13 $pf = \cos \Theta = P/VI$

Since in many cases we are interested in the energy transfer capability of an electric current, we often use an effective value for the alternating current. This effective value is found to be the "square root of the mean squared value" or the root mean square (RMS) value. The RMS value produces the same heating effect in a resistance as a direct current of the same ampere value. The RMS value is found to be:

FORMULA 1-14 $I_{RMS} = I_{PEAK} \times 0.707$

FORMULA 1-15 $V_{RMS} = V_{PEAK} \times 0.707$

The RMS values are commonly used when referring to distribution voltages and currents since this is the value measured by voltmeters and ammeters. Consequently, a rating of 115 volts for an appliance operation is the RMS value. Additionally, RMS values of voltage and current are generally used in Formula 1-12 since average power is generally of primary interest.

SIM 1-1

Using Formula 1-5 write an expression for household voltage (i.e., 115 VAC). Express frequency in terms of Hz and assume the phase angle is zero.

ANSWER

Formula 1-5 is: a = A cos (ωt + α). Since 115 VAC is an RMS value, A is 115 volts/0.707 = 163 volts. From Formula 1-6 ω = 2π f and therefore for a 60 Hz distribution frequency the expression is:

v = 163 cos (2π60t) volts

THREE-PHASE POWER

Most power is generated and transmitted in 3 phases in which three wires are utilized with the voltage in each equal in magnitude but differing in phase by 360°/3 = 120° (See Figure 1-8). Three-phase power offers the following advantages over single-phase:

1. Generators are more efficient.
2. Motors start and run smoother.
3. Power is constant rather than fluctuating during the cycle.

Phasor addition of the currents and voltages of a three-phase power system yield the following expression for power:

FORMULA 1-15   P = √3 VI cos Θ watts

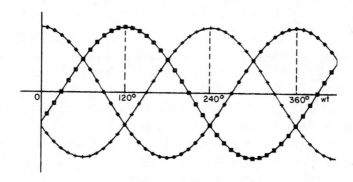

Figure 1-8. Three-Phase Voltages

TRANSFORMERS

A transformer is an electrical device which converts alternating voltages and currents from one value to another (either up or down). This is accomplished with approximately equal power transfer (i.e., if the voltage is increased, the current is decreased a proportionate amount. See Figure 1-9).

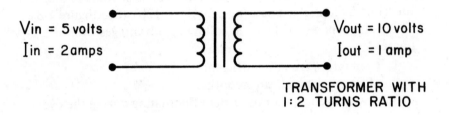

V_{in} = 5 volts
I_{in} = 2 amps

V_{out} = 10 volts
I_{out} = 1 amp

TRANSFORMER WITH
1:2 TURNS RATIO

Figure 1-9. Transformer Operation

Transformers contain two separate coils of insulated wire wound on an iron frame. Alternating current flowing through a coil develops a magnetic field that expands and contracts in step with the changes in current. The magnetic field in one coil induces current to flow in the other coil by cutting through the turns of wire.

Transformers are used in many stages in distributing power from the generating station to the user. Power is generally transferred at very high distribution voltages (several thousand volts) since the associated current is relatively low and the distribution losses are much decreased. (Remember that Power loss = i^2 R and that for a fixed value of line resistance, the lower the current the much lower the power loss.) Additionally, lower current values allow smaller wires and associated current carrying equipment to be used. Electrical substations near the point of use, consisting of banks of transformers, are then used to reduce voltages to usable levels.

DISTRIBUTION VOLTAGES

Modern electrical distribution within facilities has tended toward higher voltages for many of the same reasons as utilities have (i.e., lower costs associated with lower current carrying needs). Consequently, wiring for appliances, outlets, lights, etc. (called branch circuit wiring) has tended to be routed relatively short distances to strategically located transformer load centers which are then linked to a central distribution panelboard. This approach allows power to be delivered at the required voltage while minimizing long branch circuit runs at low voltage. There are three basic voltage distribution systems that are used today as described below (note that more than one of these systems may be present in a facility).

SINGLE-PHASE

This is a commonly used system in residential and small commercial buildings. See Figure 1-10. 240 volt branch circuits can be used for power loads such as clothes dryers, electric ranges, etc. 120 volt branch circuits are used for lighting and receptacles.

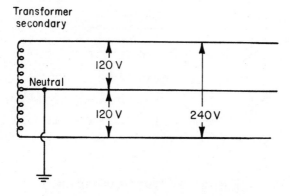

Figure 1-10. Single Phase Distribution

WYE SYSTEM

This system is designated wye because the connection of the transformer secondary coils resembles a "Y." This system provides 3-phase and single-phase power at a variety of voltages. See Figure 1-11.

The 120/208 volt configuration is generally available in all except heavy industrial facilities to some extent. The 120 volt circuits are used for receptables and lighting while the 208 volt circuits can be used for motor loads.

The 277/480 volt system is rapidly becoming the system of choice in commercial and industrial facilities because of the advantages mentioned earlier of utilizing higher distribution voltages. 480 volt, 3-phase circuits are used for motor loads; 277 volt single-phase circuits are used for fluorescent and HID (high intensity discharge) lighting; 120, 240 or 120/208 volt circuits are available from transformers for receptacles and miscellaneous loads.

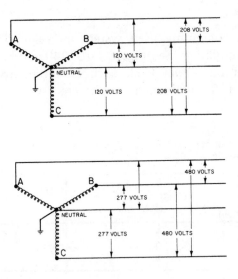

Figure 1-11. Wye Distributions

DELTA SYSTEMS

The delta-connected secondary system (Figure 1-12) is available with phase-to-phase voltages of 240, 480 or 600 volts. This system is used where motor loads represent a large part of the total load (i.e., some industrial facilities). In a typical installation, 480 volt 3-phase circuits supply motor loads while lighting and receptacle circuits are supplied by a single or 3-phase step-down transformer (as shown in the figure).

A variation on the delta connection is also shown in Figure 1-12. In this system one of the transformer secondary windings is center-tapped to obtain a grounded neutral conductor to the two phase legs to which it is connected. Motors are supplied at 240 volts, 3-phase and 120 volt single phase circuits are supplied by the neutral conductor and phases B or C.

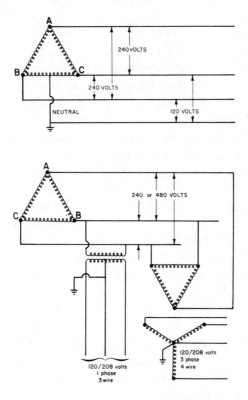

Figure 1-12. Delta Distribution

MEASURING ELECTRICAL SYSTEM PERFORMANCE

The ammeter, voltmeter, wattmeter, power factor meter, and footcandle meter are usually required to do an electrical survey or audit. These instruments are described below.

Ammeter and Volmeter

To measure electrical currents, ammeters are used. For most audits, alternating currents are measured. Ammeters used in audits are portable and are designed to be easily attached and removed.

There are many brands and styles of snap-on ammeters commonly available that can read up to 1000 amperes continuously. This range can be extended to 4000 amperes continuously for some models with an accessory step-down current transformer.

The snap-on ammeters can be either indicating or recording with a printout. After attachment, the recording ammeter can keep recording current variations for as long as a full month on one roll of recording paper. This allows studying current variations in a conductor for extended periods without constant operator attention.

The ammeter supplies a direct measurement of electrical current which is one of the parameters needed to calculate electrical energy. The second parameter required to calculate energy is voltage, and it is measured by a voltmeter.

Several types of electrical meters can read the voltage or current. A voltmeter measures the difference in electrical potential between two points in an electrical circuit.

In series with the probes are the galvanometer and a fixed resistance (which determine the voltage scale). The current through this fixed resistance circuit is then proportional to the voltage and the galvanometer deflects in proportion to the voltage.

The voltage drops measured in many instances are fairly constant and need only be performed once. If there are appreciable fluctuations, additional readings or the use of a recording voltmeter may be indicated.

Most voltages measured in practice are under 600 volts and there are many portable voltmeter/ammeter clamp-ons available for this and lower ranges.

Wattmeter and Power Factor Meter

The portable wattmeter can be used to indicate by direct reading electrical energy in watts. It can also be calculated by measuring voltage, current and the angle between them (power factor angle).

The basic wattmeter consists of three voltage probes and a snap-on current coil which feeds the wattmeter movement.

The typical operating limits are 300 kilowatts, 650 volts, and 600 amperes. It can be used on both one- and three-phase circuits.

The portable power factor meter is primarily a three-phase instrument. One of its three voltage probes is attached to each conductor phase and a snap-on jaw is placed about one of the phases. By disconnecting the wattmeter circuitry, it will directly read the power factor of the circuit to which it is attached.

It can measure power factor over a range of 1.0 leading to 1.0 lagging with "ampacities" up to 1500 amperes at 600 volts. This range covers the large bulk of the applications found in light industry and commerce.

The power factor is a basic parameter whose value must be known to calculate electric energy usage. Diagnostically it is a useful instrument to determine the sources of poor power factor in a facility.

Portable digital KWH and KW demand units are now available.

Digital units can have read-outs of energy usage in both KWH and KW demand or in dollars and cents, including instantaneous usage, accumulated usage, projected usage for a particular billing period, alarms when over-target levels are desired for usage, and control-outputs for load-shedding and cycling are possible.

Continuous displays or intermittent alternating displays are available at the touch of a button of any information needed such as the cost of operating a production machine for one shift, one hour or one week.

Footcandle Meter

Footcandle meters measure illumination in units of foot-candles through light-sensitive barrier layer of cells contained within them. They are usually pocket size and portable and are meant to be used as field instruments to survey levels of illumination. Footcandle meters differ from conventional photographic lightmeters in that they are color and cosine corrected.

Chapter 2

USING THE LANGUAGE
OF THE ELECTRICAL ENGINEER

To design the electrical portions of an industrial plant or a commercial building requires knowledge of power, lighting, and control. The viewpoint in the following chapters is that of an electrical engineer, designing a new facility. This "role playing" experience will enable the reader to gain a better understanding of the elements that go into the design, to deal better with contractors and in-house designers, and to interpret drawings of an existing facility.

OBJECTIVES OF ELECTRICAL DESIGN

Three elements usually comprise the basis of electrical design; namely, technical proficiency, cost considerations, and overall schedules.

- *Technical Proficiency.* The electrical design should meet the facility's requirements, local and national codes and all safety requirements.

- *Cost Considerations.* Decisions relating to materials of construction, and first and operating costs should be analyzed using the principles of life-cycle costing. Electrical engineering and design manpower requirements should be established, monitored and controlled.

- *Schedules.* Schedules should be made for all engineering and construction activities. Delays in engineering or construction activities can be very costly. Monitoring progress, spotting areas of concern and implementing corrective action is required to ensure an orderly design.

ACTIVITIES OF THE ELECTRICAL
ENGINEER/DESIGNER

Throughout this book you will be involved in the design of a hypothetical facility. Since a facility can contain process, power generation, and office areas you will gain a broad exposure to electrical problems.

It will also be seen that to design the electrical portions of a facility requires both an engineering and a design approach.

Typical Problem

The Ajax Company* is building a plant. The first step for the electrical engineer/designer is to help the client (plant) establish their needs. Many clients know what they want, but they need help in defining what has to be done. The design engineer studies all aspects of the client's requirements and establishes the design criteria. He must consider, for example:

• How to service the loads of the plant. This includes determining the voltage level to best service the load economically; should overhead or underground distribution be used; negotiations with the utility company to establish system requirements; and the type of reliability required.

• Type of lighting system required.

• Auxiliary systems required.

• Type of equipment required.

• Control considerations.

• Economic considerations.

Once the design criteria have been established the production of the job can begin. Design drawings and specifications to meet the job criteria can be made.

Engineering Activities

Using the above design criteria, the engineering activities include:

• Establishing criteria for One Line Diagram.

---

*The Ajax Company is fictitious, but the principles you will experience are not.

• Establishing budgets, schedules and manpower requirements.

• Writing specifications.

• Inspecting equipment.

• Coordinating activities of design, vendors, subcontractors and client.

• Checking vendor prints.

• Performing special studies.

• Preparing estimates.

Design Activities

Design activities are involved in the preparation of the various drawings required to convey installation information. Typical drawings include:

• *One Line or Single Line Diagram.* Figure 2-1 illustrates a typical single line diagram. This diagram is a simple schematic which identifies how power is distributed from the source to the user. Equipment such as switchgear, substations, motor control centers and motors are illustrated. The diagram also indicates the voltage levels, bus capacities, fuse or breaker ratings, key metering and relaying and other identification which will aid in describing the electrical distribution. Depending on the size of the system, sometimes several one line diagrams are needed. A main one line diagram illustrates the primary switchgear and substations. Motor Control Center one lines can then be used to show all motors and how they are being fed.

• *Power Plans.* Figure 2-2 illustrates a typical power plan. This diagram is a physical plan which is drawn to scale. It shows where all motors and loads are located and how they are fed. Conduit and cable sizes are indicated (if the project is large, conduit and cable sizes are indicated on separate sheets with only the description appearing on the power plan).

• *Elementary Diagrams.* Figure 2-3 illustrates a typical elementary diagram. This is a schematic which indicates how a system is controlled. Typical control devices such as pushbuttons, limit switches, level switches and pressure switches are used to energize relays, motor holding coils and solenoid valves.

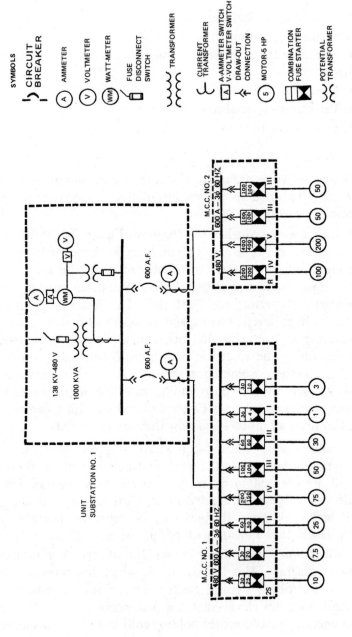

Figure 2-1. Typical One-Line Diagram

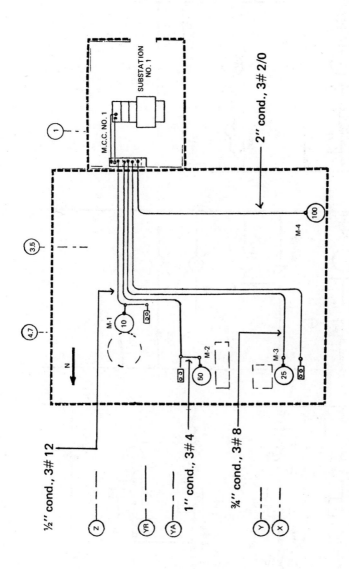

Figure 2-2. Typical Power Plan

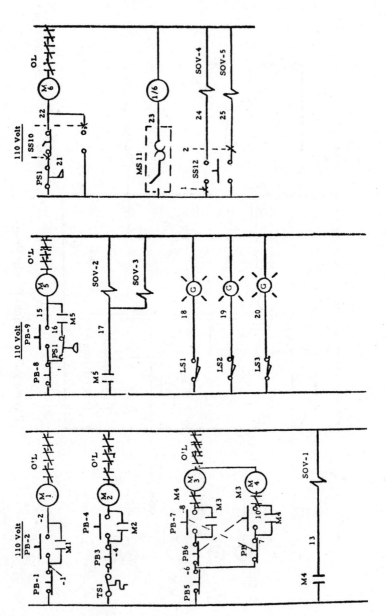

Figure 2-3. Typical Elementary Diagram

The elementary diagram indicates how a system operates, but not the physical properties of each element.

• *Interconnection Diagram.* A typical interconnection diagram is illustrated in Figure 2-4. The elementary diagram is used as the basis for this drawing. All relays are shown in their relative location. Terminal numbers and point-to-point wiring are shown. This drawing is used by the electrician to connect the wires to each terminal. Sometimes the information contained on the diagram is summarized on a schedule by a computer, thus eliminating the need for this drawing.

• *Lighting Plans.* A typical lighting plan is illustrated in Figure 2-5. This diagram is a physical plan which is drawn to scale. It shows the location of fixtures, outlets, and lamp circuiting. All lighting fixtures are either identified by a symbol on the lighting drawing or in a separate symbol list. Lighting panelboards may also be shown on this plan if space allows, or on separate schedule sheets. See Figure 2-6. A lighting panelboard schedule shows the number, location and power consumed by the lights on each branch circuit. Additionally, the circuit breaker sizes and the phase to which each is connected is shown.

• *Details and Miscellaneous Diagrams.* Detail drawings may be used to supplement the power or lighting plans. These details may indicate a blow-up for field fabrication, a special support, an electrical room, outdoor substation layout, or any other item which needs further clarification.

Miscellaneous drawings may also be required to summarize materials, indicate instrument locations, or cover any drawing necessary to describe the complete electrical system.

• *Grounding Drawings.* To provide safety to personnel, a facility must be adequately grounded. A grounding drawing usually indicates the main grounding loop around the plant and typical details for grounding steel columns, motors, tanks, etc.

• *Variations.* Figures 2-1 through 2-6 may be combined as long as the information contained is clearly depicted.

Several categories may be eliminated by using visual aids such as a model. Power plans which show motor runs can be eliminated if all conduit runs are shown on the model.

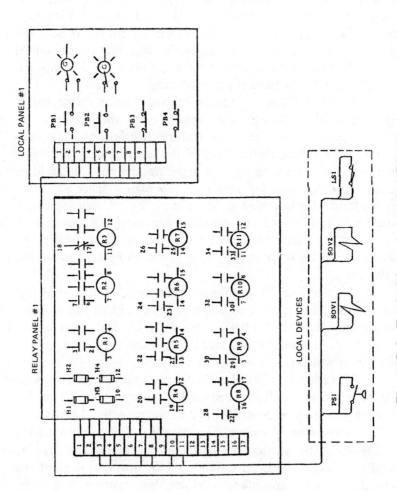

Figure 2-4. Typical Interconnection Diagram

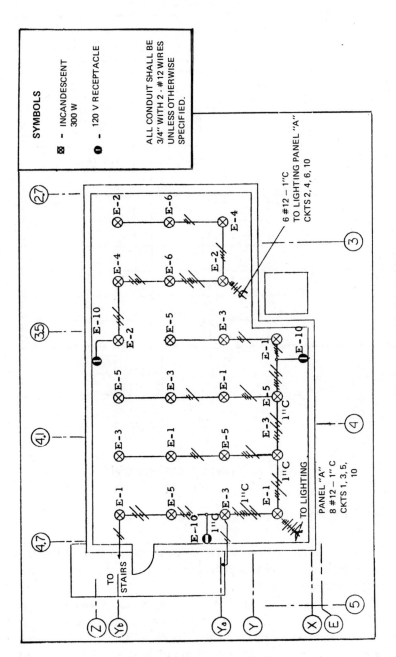

Figure 2-5. Typical Lighting Plan

LIGHTING PANEL "A"

CIRCUIT NO.	SERVICE	NO. OF OUTLETS	WATTS	NEUTRAL A B C	WATTS	NO. OF OUTLETS	SERVICE	CIRCUIT NO.
1	WAREHOUSE	5	1500		900	3	WAREHOUSE	2
3	WAREHOUSE	5	1500		600	2	WAREHOUSE	4
5	WAREHOUSE	4	1500		600	2	WAREHOUSE	6
7	H&V UNITS	-	320		990	11	WAREHOUSE STAIRS	8
9	BLANK	-	-		600	3	RECEPTACLE	10
11	BLANK		-		1200		SPARE	12
13								14
15								16
17								18
19								20
21								22
23								24

12 CIRCUIT PANEL
20 AMPERE BREAKER
120/208 V – 3 PHASE
4 WIRE

PANEL LOADING	
PHASE	WATTS
A	3710
B	2700
C	2100
CONNECTED LOAD	8510
SPARES	1200
TOTAL	9710

Figure 2-6. Typical Lighting Panel

DESIGN ACTIVITY MANHOURS

When dealing with outside contractors and evaluating the scope of a project it is useful to have an understanding of how manhour estimates are made.

Table 2-1 summarizes the drawing types, drawings sizes, and the scales to which they are commonly drawn, plus the manhours required to produce the drawing. The hours for each drawing depend on the amount of detail shown and the type of firm that is making the drawing. For instance, an architectural firm designing a commercial lighting project may just choose the fixture type and design a layout. A designer in an engineering firm would probably show in addition to the above, the circuits to each fixture and a lighting panelboard schedule. The more detail, the more hours. This table should serve as a guide. The practices in each firm and past job performances will be the prevalent factors in determining the drawing details and the estimated manhours.

Table 2-1. Typical Drawing Requirements

DRAWING	*SCALE*	*HOURS TO DESIGN DRAWING (including calculations)*	*TYPICAL DRAWING SIZE (Inches)*
Lighting	1/8"=1'	30-50	30 x 42 24 x 36
Power	1/4"=1'	50-75	30 x 42 24 x 36
Lighting Schedules	None	10	9 x 12
Conduits & Cable Lists	None	10	12 x 18
Elementary Wiring	None	75-100	30 x 42 24 x 36
*One Lines	None	75-100	30 x 42 24 x 36
Interconnection	None	30-50	30 x 42 24 x 36
Grounding Drawings	1"=100' (Depends on Plot Plan)	30-40	30 x 42 24 x 36
Miscellaneous Details	Depends on Detail	50-100 Depends on Detail	30 x 42 24 x 36 9 x 12

*A unit substation and fifty motors can usually be fitted on a full-size one line diagram.

In the following pages you will experience job situations. Each simulation experience will be denoted by SIM. The answer will be written below the problem. Cover the answer so that you can play the game.

SIM 2-1

Estimate the number of power and lighting layouts for the following details.

Client: Ajax manufacturing plant "A"
             Basement — 50' x 200'
             Operating Floor 50' x 200'

Answer

From Table 2-1
Power Plan — 1/4"=1'
Thus 50' x 200' will fit on a 12" x 50" drawing.
Since a standard size drawing is 30" x 42", half of the operating and basement floor can fit on one drawing. Two drawings are required.

Lighting Plan — 1/8"=1'
$$\frac{50}{8} \times \frac{200}{8} = 6'' \times 25''$$

One drawing will be sufficient. 30" x 42".

SIM 2-2

Estimate the number of one line diagrams for the plant of SIM 2-1.

Given: 28 motors on two motor control centers with an estimated load of 800 KVA. Assume one substation will feed load.

Answer

Based on the above load, one unit substation and the associated Motor Control Centers will fit on one drawing.

SIM 2-3

Estimate the number of elementary and interconnection drawings for SIM 2-2. Each motor needs a separate stop-start control scheme. Control schemes should be provided so that 14 solenoid valves can be activated. All relays, pushbuttons, etc., are located on a local panel. Assume 100 elementary lines per drawing. Allow 2 spaces between each scheme

Answer

Estimating elementary and control schemes is a very difficult task. Usually at the beginning of the project it is difficult to get an exact description of the control.

Assume stop-start scheme. From Figure 2-3 two lines per scheme; thus 28 x 2 = 56 lines. Assume 2 spaces between each motor.

Scheme: 28 x 2 = 56 lines

Assume 2 lines per each solenoid scheme
28 x 2 = 56 lines

Since more than 100 lines are required, two drawings are estimated.

In estimating the interconnection diagrams, assume an interconnection diagram is needed for each motor control center and local panel. By actually counting the number of terminals required, the number of drawings could be reduced later on.

SIM 2-4

Compile a drawing list with estimated manhours for problems SIM 2-1 through 2-3.

Answer

DRAWING NO.	DESCRIPTION	ESTIMATED MANHOURS
DWG 1	One Line Diagram	75
DWG 2	Power Plan	50
DWG 3	Power Plan	50
DWG 4	Lighting Diagram—Basement Operating	30
DWG 5	Grounding Drawing	30
DWG 6	Conduit and Cable Schedule Assume 2*	20
DWG 7	Lighting Schedule—Assume 2	20
DWG 8	Elementary Diagram	75
DWG 9	Elementary Diagram	75
DWG 10	Interconnection MCC No. 1	30
DWG 11	Interconnection MCC No. 2	30
DWG 12	Interconnection Local Devices	30
	TOTAL MANHOURS	515

*One drawing for Basement and one drawing for Operating Floor.

ENGINEERING ACTIVITY MANHOURS

It is more difficult to evaluate the engineering activities at the beginning of the project because many of the activities are involved with intangibles such as coordination.

From the details of the design an estimate can be made of the number of requisitions which are needed to purchase equipment.

• To write a specification using a previous one as a guide may take from 10 to 15 hours.

• To write a completely new specification may take 60 hours.

• Compiling material requisitions for quotation and purchase, evaluating sellers' quotes, and checking vendor prints could vary from 40 to 100 hours per requisition. The manhours required depends on the complexity of the equipment being purchased.

Many times engineering activities may be estimated as a percentage of design manhours. Coordination and general engineering activities may be from 10-20% for small projects below 5000 hours and 15-30% for larger ones which need more coordination. Specifications, requisitions and special studies should be added in separately.

SIM 2-5

Estimate the engineering hours for the design of 500 manhours as indicated in SIM 2-4. Assume no special studies; past specifications are available and minimum coordination time is required.

Answer

Substation Specification	10
Requisition and Vendor Prints	40
Motor Control Centers (2)	
Specifications	10
Requisition and Vendor Prints	40
Coordination—10% (500)	<u>50</u>
	150 manhours

Notice that the engineering activities on an industrial project are usually only a fraction of the total time required for design.

TYPES OF DESIGN ORGANIZATIONS

The design section of the facilities department of a plant or that of an outside consulting firm is usually organized either by departments or by tasks. The two organizations widely encountered are:

• *Department Oriented.* In the department oriented firm, each group (Electrical, Structural, Civil, Architectural, HVAC, Piping, and Plumbing) is separate from the other. The Department Head usually has his people grouped together. All information usually channels down from the Department Head.

• *Task Force.* Each Department Head chooses people for a project. These people move out of the department and operate as a team. Each individual usually gains responsibility since the departmental chain of command has disappeared.

In the "Project" or "Task Force" approach, a Project Engineer is assigned to the job to help coordinate the various disciplines.

KNOW THY VENDOR

Manufacturers provide many services in order to sell their products. Good relations with manufacturers can aid in electrical design.

Typical services include:

• Computer analysis for determining the most economical lighting system.

• Technical brochures.

• Cost information.

• Proven expertise in their field.

Since most of these services are free, good communications with vendors is extremely important. Remember vendors are offering these services as a means of selling their product. Be careful not to get "married" to one vendor, and to look objectively at all information he is offering.

ELECTRICAL SCHEDULE

Always Last

Electrical design cannot proceed without the motor horsepowers. For example, the heating and ventilating group must size their fans, the mechanical group must select their pumps, and the architectural group must select the automatic roll-up doors before all motor horsepower can be given to the electrical engineer. The electrical engineer is vulnerable to any changes from the other departments since their changes will probably affect the electric load.

Other aspects which affect design are firm equipment locations. This input is required to design the power and lighting drawings. A description of operation or a logic diagram is needed before the elementary and interconnection drawings can be designed. Thus the electrical design is usually the last to be finished on a project. Because of this, the electrical engineer is always under pressure to complete the design.

Critical Path

In many cases it is the electrical group which determines the critical path. Remember that the delivery of electrical equipment such as switchgear or high voltage bus duct may take up to a year to fabricate. Thus, one of the first activities the electrical engineer should do is purchase equipment. This means that many times an estimate of electrical loads must be used to purchase equipment in order to meet the schedules.

A typical schedule is illustrated in Figure 2-7.

JOB SIMULATION SUMMARY PROBLEM

JOB 1

Background

At the end of several chapters, a job simulation summary problem will be given. In these problems you will play the role of an Electrical Engineer working on Process Plant 2 of a grass roots (new) project for the Ajax Corporation. The plant is comprised of two identical modules. The response to each of these problems will be needed in order to complete the subsequent chapters.

The first task will be to identify the electrical loads from an equipment list which was given to the engineer by the plant. From this list indicate which loads require electric power and from which group you would expect to receive the information.

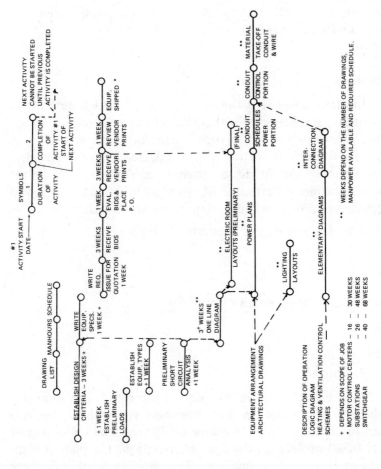

Figure 2-7. Schedule of Electrical Activities for Plant Design

For example:

A — Architectural
M — Mechanical
H — Heating and Ventilating

Equipment List — Module #1

Note: Module #2 is identical.

D-1	Tank #1
AG-1	Agitator for Tank #1
H-2	Heat Exchanger
CF-3	Centrifuge
FP-4	Feed Pump
TP-5	Transfer Pump
CTP-6	Cooling Tower Feed Pump
D-7	Water Chest
V-8	Vessel #1
CT-9	Cooling Tower
HF-10	H&V Supply Fan
HF-11	H&V Exhaust Fan
UH-12	Unit Heater
BC-13	Brine Compressor
A-14	Air Operated Motor
T-15	Turbine #1
C-16	Conveyor
H-17	Hoist
ES-18	Exhaust Stack
SC-19	Self-cleaning Strainer
VS-20	Vacuum Separator
PO-21	Pneumatic Oscillator
RD-22	Roll-up Door

The plant has also indicated that one unit substation will handle the load and that each module should be fed from a motor control center.

The area of the Process Plant 2 is as follows:

Basement	20 ft x 400 ft
Operating Floor	20 ft x 400 ft

From the above information prepare a drawing list and submit the cost to do this project. Assume the project is complex (use higher hours per drawing) and that the unit rate is $60 per technical manhour which includes overhead, profit and fees.

Analysis

Once it is known which equipment requires a motor, it is necessary to obtain the electrical loads from each discipline and compile a motor list. A completed motor list for each module would look as follows:

EQUIP. NO.	NAME	HP	TYPE	DESIGN DISCIPLINE
AG-1	Agitator	60		M
CF-3	Centrifuge	100	Reversing	M
FP-4	Feed Pump	30		M
TP-5	Transfer Pump	10		M
CTP-6	Cooling Tower Feed Pump	25		M
CT-9	Cooling Tower	20	2-Speed	M
HF-10	H&V Supply Fan	40		H
HF-11	H&V Exhaust Fan	20		H
UH-12	Unit Heater	1/6		H
BC-13	Brine Compressor	50		M
C-16	Conveyor	20	Reversing	M
H-17	Hoist	5	Local starter by vendor	M
SC-19	Self-cleaning Strainer	3/4		M
RD-22	Roll-up Door	1/8	Local starter by vendor	A

Next, prepare a drawing list and estimate the manhours required to produce the drawings and specifications.

Typical Drawing List and Manpower Estimate

DESIGN		
DRAWING NO.	DESCRIPTION	ESTIMATED MANHOURS
DWG 1	One Line Diagram	100
DWG 2	Power Plan — Basement	75
DWG 3	Power Plan — Operating Floor	75
DWG 4	Lighting Diagram — Basement and Operating Floor	50
DWG 5	Grounding Drawing	40
DWG 6	Conduit & Cable Schedule (2)	20
DWG 7	Lighting Schedule (2)	20
DWG 8	Elementary Diagram	100
DWG 9	Elementary Diagram	100
DWG 10	Interconnection MCC No. 1 and Local Device	50
DWG 11	Interconnection MCC No. 2 and Local Device	50
		680 Hrs

ENGINEERING		
Substation Specification		15
Requisition and Vendor Prints		100
Motor Control Centers		15
Specifications, Requisitions, Vendor Prints		100
Engineering Coordination — 20% (680)		130
	Total	360

TOTAL HOURS	1040 — say 1100 Hrs	
TOTAL COST	$66,000	

The $66,000 cost is probably on the high side since the two modules are identical and the time to do the second module would be less. Also the assumption that the job is complex does not seem reasonable, based on the scope just outlined.

As you can see, estimating is an "art" rather than an exact science.

Chapter 3

EQUIPMENT SELECTION
CONSIDERATIONS

The electrical engineer is responsible for distributing power from the service entrance to its varied destinations throughout the facility. This chapter introduces some of the equipment that the engineer must size and specify when designing a safe and efficient power distribution system. Motor control center layout as well as circuit breaker selection considerations are presented in this chapter.

ELECTRICAL EQUIPMENT

Electrical equipment commonly specified is as follows:

• *Switchgear-Breakers*—used to distribute power and provide overcurrent protection for high voltage applications.

• *Unit Substation*—used to step down voltage. Consists of a high voltage disconnect switch, transformer and low-voltage breakers. Typical transformer sizes are 300 KVA, 500 KVA, 750 KVA, 1000 KVA, 2000 KVA, 2500 KVA and 3000 KVA.

• *Motor Control Center (M.C.C.)*—a structure which houses starters and circuit breakers or fuses for motor control. It consists of the following:

    (1) Thermal overload relays which guard against motor overloads;

    (2) Fuse disconnect switches or breakers which protect the cable and motor and can be used as a disconnecting means;

    (3) Contactors (relays) whose contacts are capable of opening and closing the power source to the motor.

• *Panelboard/Switchboard* — Breakers; used to distribute power and provide overcurrent protection to motor control centers, lighting, receptacles and miscellaneous power circuitry within a building.

MOTOR CONTROL CENTER BREAKERS AND FUSES

It should be noted that substation breakers are different from switchgear breakers, motor control center breakers and panelboard breakers. Switchgear breakers may be of the "vacuum type," whereas substation breakers may be of the "magnetic air circuit type" and motor control center and panelboard breakers may be of the "molded case" type.

One of the most important ratings for a circuit breaker is its ability to interrupt a short circuit (also called a fault). When a short circuit occurs, many thousand amperes of peak (or asymmetrical) current can flow which can "fuse" the contacts of a circuit breaker closed, thereby preventing it from operating to prevent fires, explosions, etc. The interrupting capacity (see Table 3-1) rating of a circuit breaker determines its ability to "clear" a fault of a certain magnitude.

Note, that low voltage (less than 600 volts) circuit breakers often have two trip ranges. The first or "thermal" protection is adjusted to prevent the wiring from overheating when overloaded for a period of time. The second, instantaneous trip, is often adjustable and is used to quickly interrupt a circuit should the high levels of current associated with a short circuit occur.

Table 3-1 summarizes breaker and starter sizes for motor control centers. Table 3-2 summarizes dual element fuse and switch sizes for motor control centers. The fuse and breaker sizes indicated in these tables are based on vendor's data. As long as the values are below specified values listed in the National Electrical Code and coordinate with the motor, the selection is satisfactory.

There are several types of fuses commonly used. Each type is characterized by its time to isolate the fault, interrupting rating, and current limiting property. Fuse types include: standard fuses, time delay fuses, current limiting fuses, and dual element fuses.

Table 3-1. Combination Breaker-Starter Ratings

MOTOR HP	STARTER SIZE	BREAKER TRIP* Amps	BREAKER FRAME** Amps	M.C.C. SPACE†
1	—	15	100	14"
1½	—	15	100	14"
2	—	15	100	14"
3	—	15	100	14"
5	—	15	100	14"
7½	—	30	100	14"
10	—	40	100	14"
15	II	50	100	14"
20	II	50	100	14"
25	II	50	100	14"
30	III	70	100	14"
40	III	100	100	28"
50	III	100	100	28"
60	IV	125	225	28"
75	IV	150	225	28"
100	IV	200	225	28"
125	V	225	225	42"
150	V	300	400	42"
200	V	350	400	42"

BREAKER TYPE	TRIP RANGE	ASYM. AMPS
FA	15-100	15,000
JA	70-225	20,000
KA	70-225	25,000
LA	125-400	35,000

* Check chosen vendor for specific recommendations.

** Minimum size — check short-circuit rating.

† Based on M.C.C. Vendor's data for FVNR (Full Voltage Non-Reversing Starters). Check chosen vendor for specific details.

Table 3-2. Combination Fuse-Switch Ampere Ratings

MOTOR HP	460 V F.L.A.	FUSE*	SWITCH	M.C.C.** SPACE
1	1.8	4	30	14"
1½	2.6	5	30	14"
2	3.4	8	30	14"
3	4.8	10	30	14"
5	7.6	15	30	14"
7½	11	20	30	14"
10	14	25	30	14"
15	21	30	30	14"
20	27	40	60	14"
25	34	50	60	14"
30	40	60	60	28"
40	52	80	100	28"
50	65	100	100	28"
60	77	125	200	42"
75	96	150	200	42"
100	124	200	200	42"
125	156	250	400	70"
150	180	300	400	70"
200	240	400	400	70"

*Based on Dual Element Fuses.

**Based on M.C.C. Vendor's data for FVNR Starters.
Check chosen vendor for specific details.

Advantages of Fuses Over Breakers

The advantages of fuses over breakers are:

• Higher interrupting ratings (100,000 amps-dual element fuses).

• Current limiting action—will limit the short circuit current downstream of fuse.

• Lower cost.

• Less affected by corrosive atmosphere.

• Less affected by moisture.

• Less affected by dust.

Advantages of Breakers Over Fuses

The advantages of breakers over fuses are:

- Resetable.
- Electrically operated breakers can be remotely operated.
- Adjustable characteristics.

SIM 3-1

Indicate the starter size for 10, 30 and 100 HP motors.

Answer (from Table 3-1)

10 HP	Size I
30 HP	Size III
100 HP	Size IV

SIM 3-2

The short-circuit current available at a motor control center is 15,000 amperes asymmetrical. Indicate the frame (continuous rating) and trip (current at which breaker will open) sizes for 7½, 30, 60 and 100 HP motors.

Answer (from Table 3-1)

Horsepower	Trip Size	Frame Size
7½	30	100
30	70	100
60	125	225
100	200	225

SIM 3-3

The short-circuit current available at a motor control center is 25,000 amperes asymmetrical.

Repeat SIM 3-2.

Answer (from Table 3-1)

Horsepower	Trip Size	Frame Size	
7½	70	225	Min. Size
30	70	225	Breaker
60	125	225	KA−25,000
100	200	225	

Note: For large short-circuit currents, breakers are impractical as indicated above. Either the short-circuit current should be decreased or fuses should be used instead of breakers.

SIM 3-4

Indicate the fuse and switch sizes for the following: 3, 10, 50, 75 HP motors.

Answer (from Table 3-2)

Horsepower	Fuse Size	Switch Size
3	10	30
10	25	30
50	100	100
75	150	200

MOTOR CONTROL CENTER LAYOUTS

Figure 3-1 shows a typical outline for a M.C.C. Dimensions vary between vendors and the space allocated for wiring depends on whether cables enter from top or bottom and if terminal blocks are required in upper or lower section.

Usually terminal blocks are located in the individual starter cubicle and the top or bottom is used only for wiring between sections.

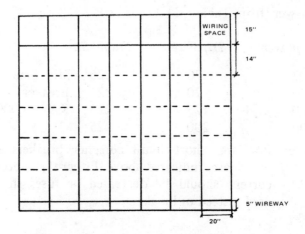

Figure 3-1. M.C.C. Outline
(Consult Vendor for Specific Dimensions)

Referring to Figure 3-1, a vertical section is a structure with an overall height of 90″, a width of 20″ and a depth of nominally 13″-20″. It includes a horizontal feeder bus at the top and a vertical bus bar to accept the plug-in motor and starter units. A unit is the motor starter-disconnect module that fits into the vertical section and is covered by a door. Thus, the motor control center consists of vertical sections bolted together and connected with a common horizontal feeder bus bar. The working height is the available space in any section for motor control center units.

Layout

When designing a motor control center keep in mind the following:
- Larger units should be placed near the bottom of a section for easier maintenance.
- Place the various required units in as many sections as necessary to accommodate them.

- All units fit into the vertical section merely by moving the unit support brackets to fit the structure. Filler plates can be used for leftover spaces.
- The commonly used sizes which are used in conjunction with Figure 3-1 are summarized in Figure 3-2. The sizes shown in Figure 3-2 are based on combination fuse starters.

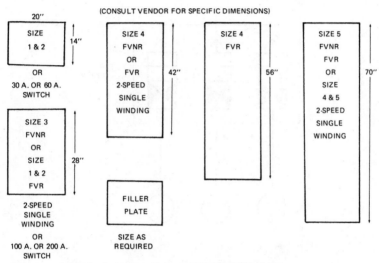

NOTES: 1. FVNR REFERS TO FULL VOLTAGE NONREVERSING STARTERS
2. FVR REFERS TO FULL VOLTAGE REVERSING STARTERS

Figure 3-2.
(Consult Vendor for Specific Dimensions)

Note: The minimum bus for a motor control center is 600 amperes. Initial sizing is usually based on 400 to 500 HP per motor control center. This is usually adequate. A detailed load check should be made when the design is firm.

M.C.C. One-Line Diagram

Typical symbols used for a M.C.C. one-line diagram are illustrated in Figure 3-3.

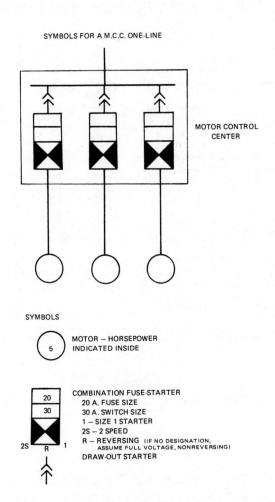

SYMBOLS FOR A M.C.C. ONE-LINE

MOTOR CONTROL CENTER

SYMBOLS

MOTOR — HORSEPOWER INDICATED INSIDE

COMBINATION FUSE-STARTER
20 A. FUSE SIZE
30 A. SWITCH SIZE
1 — SIZE 1 STARTER
2S — 2 SPEED
R — REVERSING (IF NO DESIGNATION, ASSUME FULL VOLTAGE, NONREVERSING)
DRAW-OUT STARTER

Figure 3-3. Symbols for a M.C.C. One-Line Diagram

MOTORS

• *Squirrel Cage Induction Motors* are commonly used. Three phase versions of motors require three power leads. For two-speed applications several different types of motors are available. Depending on the process requirements such as constant horsepower or constant torque, the windings of the motor are connected differently. The theory of two-speed operation is based on Formula 3-1.

(Formula 3-1)

$$\text{Frequency} = \frac{\text{No. of poles X speed}}{120}$$

Thus, if the frequency is fixed, the effective number of motor poles should be changed to change the speed. This can be accomplished by the manner in which the windings are connected. Two-speed motors require six power leads.

• *D. C. Motors* are used where speed control is essential. The speed of a D. C. Motor is changed by varying the field voltage through a rheostat or through higher efficiency solid state controls. A D.C. Motor requires two power wires to the armature and two smaller cables for the field.

• *Synchronour Motors* are used when constant speed operation is essential. Synchronous motors are sometimes cheaper in the large horsepower categories when slow speed operation is required. Synchronous motors also are considered for power factor correction. A .8 P.F. synchronous motor will supply corrective KVARs to the system. A synchronous motor requires A. C. for power and D. C. for the field. Since many synchronous motors are self-excited, only the power cables are required to the motor.

• *Energy Efficient* versions of induction motors are available which can offer significant energy savings depending on the application. Additionally, these motors offer significant improvement in power factor. See Figures 3-4 and 3-5.

• *Variable Speed Drives* utilizing inverters are being used to change the frequency of power, and thereby the speed, supplied to standard induction motors. Since some motor applications, primarily for pumps and fans, can utilize reduced speed operation for much of the operating time, significant energy savings are possible. (Note, that a 50% reduction in speed yields an 88% reduction in power required for a fan.)

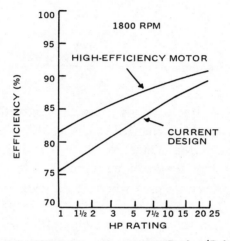

Figure 3-4. Efficiency vs Horsepower Rating (Dripproof Motors)

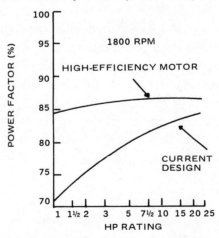

Figure 3-5. Power Factor vs Horsepower Rating (Dripproof Motors)

MOTOR HORSEPOWER

The standard power rating of a motor is referred to as a horsepower. In order to relate the motor horsepower to a kilowatt (KW) multiply the horsepower by .746 (Conversion Factor) and divide by the motor efficiency.

(Formula 3-2)
$$KW = \frac{HP \times .746}{\eta}$$

HP = Motor Horsepower
η = Efficiency of Motor

Motor efficiencies and power factors vary with load. Typical values are shown in Table 3-3. Values are based on totally enclosed fan-cooled motors (TEFC) running at 1800 RPM "T" frame.

Table 3-3.

HP RANGE	3-30	40-100
η % at		
½ Load	83.3	89.2
¾ Load	85.8	90.7
Full Load	86.2	90.9
P.F. at		
½ Load	70.1	79.2
¾ Load	79.2	85.4
Full Load	83.5	87.4

MOTOR VOLTAGES

For fractional motors, 1/3 HP and below 115 volt, single phase is used. These motors are usually fed from a lighting panel and do not appear on the M.C.C. one-line diagram. A local starter consisting of a switch and over-load element is all that is usually required.

Motors 1/2 HP to 250 HP are usually fed from a 480-volt, 3-phase, motor control center or equivalent.

Motors 300 HP and above are usually fed at 2300 or 4160 volts. The reason for this is mostly economics, i.e., price of motor, starter, cable and transformer.

SIM 3-5

Determine the number of poles for a 3600 RPM motor— 60 cycle service.

Answer

$$\text{No. of poles} = \frac{\text{Frequency X } 120}{\text{Speed}} = \frac{60 \text{ X } 120}{3600} = 2$$

SIM 3-6

Next to each motor indicate the probable voltage rating and whether it is a single- or three-phase motor. Motor HP are: 1/4, 25, 150, 400.

Answer

MOTOR HP	VOLTAGE	NUMBER OF PHASES
1/4	110 V	1ϕ
25	460 V	3ϕ
150	460 V	3ϕ
500	2300 or 4.16 KV	3ϕ

SUMMARY

Circuit breakers and fuses are used to protect wiring and equipment from fault currents. When a group of motors are being supplied with power, the circuit breakers and/or fuses are mounted in a motor control center.

Motor efficiency and power factor are dependent on size and load. Energy saving motors offer significant improvement in both efficiency and power factor.

JOB SIMULATION–SUMMARY PROBLEM

JOB 2

Background

(a) From the motor list established in Job 1 at the end of Chapter 2, indicate the rated voltage of each motor and if it is a single or three phase.

(b) The electrical engineer has been asked by the client to issue a requisition for direct purchase of the motor control centers. Assume a 30 Amp switch in each M.C.C. to take care of lighting loads and fractional horsepower motors. Make a sketch of the proposed layout and a M.C.C. one-line diagram. Assume each module is fed from a separate M.C.C. Use the horsepower and data of Job 2a, Chapter 3. Typical forms are illustrated.

Forms for Job 2b

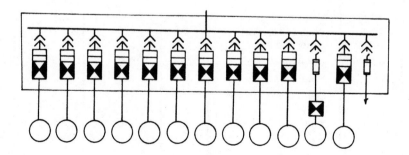

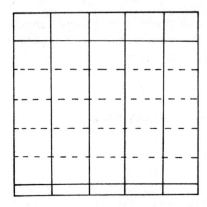

Analysis

(a) The client reviewed the motor list and expected to see the following:

Motor List — Module 1

MOTOR NO.	DESCRIPTION	HP	VOLTAGE	PHASE
AG-1	Agitator Motor	60	460	3
CF-3	Centrifuge Motor	100	460	3
FP-4	Feed Pump Motor	30	460	3
TP-5	Transfer Pump Motor	10	460	3
CTP-6	Cooling Tower Feed Pump Motor	25	460	3
CT-9	Cooling Tower Motor	20	460	3
HF-10	H&V Supply Fan Motor	40	460	3
HF-11	H&V Exhaust Fan Motor	20	460	3
UH-12	Unit Heater Motor	1/6	110	1
BC-13	Brine Compressor Motor	50	460	3
C-16	Conveyor Motor	20	460	3
H-17	Hoist Motor	5	460	3
SC-19	Self-Cleaning Strainer Motor	3/4	460	3
RD-22	Roll-Up Door Motor	1/8	110	1

JOB 2b:

A typical response the client would expect to Job 2b would look as follows:

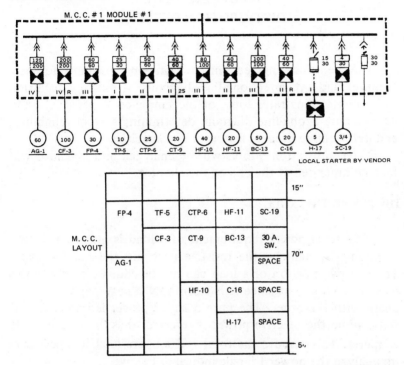

The quote from the M.C.C. vendor, based on the requisition is as follows:

Dear Sir:

The total cost for the Motor Control Center you outlined in your inquiry is $12,500 per Motor Control Center. For two Motor Control Centers the total cost is $24,000.

Delivery can be made in 16 weeks after drawing approval.

Very truly yours,

/s/ Marvin C. Cole

Chapter 4

ANALYZING POWER
DISTRIBUTION SYSTEMS

The electrical engineer initiates one-line diagrams for new facilities and interprets one-line diagrams of existing facilities. This chapter illustrates some of the simple concepts involved in extablishing a one-line diagram, determining system reliability, and determining breaker interrupting capacities

Three substation quotes will be analyzed and a vendor will be recommended at the end of the chapter.

THE POWER TRIANGLE

The total power requirement of a load is made up of two components: namely, the resistive part and the reactive part. The resistive portion of a load can not be added directly to the reactive component since it is essentially ninety degrees out of phase with the other. The pure resistive power is known as the watt, while the reactive power is referred to as the reactive volt amperes. To compute the total volt ampere load it is necessary to analyze the power triangle indicated below:

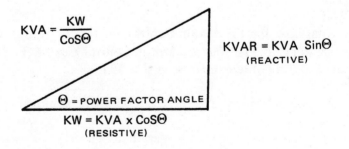

$$KVA = \frac{KW}{Cos\Theta}$$

$$KVAR = KVA \ Sin\Theta$$
(REACTIVE)

Θ = POWER FACTOR ANGLE

$$KW = KVA \times Cos\Theta$$
(RESISTIVE)

K = 1000

W= Watts

VA = Volt Amperes

VAR = Volt Amperes Reactive

Θ = Angle Between KVA and KW

CoSΘ = Power Factor

$$\tan \Theta = \frac{KVAR}{KW}$$

IMPORTANCE OF POWER FACTOR

Transformer size is based on KVA. The closer Θ equals 0° or power factor approaches unity, the smaller the KVA. Many times utility companies have a power factor clause in their contract with the customer. The statement usually causes the customer to pay an additional power rate if the power factor of the plant deviates substantially from unity. The utility company wishes to maximize the efficiency of their transformers and associated equipment. Similarly, a facility wishes to maximize the efficiency of its distribution equipment by minimizing current levels with a high power factor.

Power Factor Correction

The problem facing the electrical engineer is to determine the power factor of the plant and to install equipment such as capacitor banks or synchronous motors such that the overall power factor will meet the utility company's objectives.

Capacitor banks lower the total reactive KVAR by the value of the capacitors installed.

SIM 4-1

A total motor horsepower load of 854 is made up of motors ranging from 40-100 horsepower. Calculate the connected KVA. Refer to Table 3-3 for efficiency and power factor values.

Answer

$$KVA = \frac{HP \times .746}{Motor\ Eff.\ \times Motor\ P.F.}$$

From Table 3-3 at full load

Eff. = .909 and P.F. = .87

$$KVA = \frac{854 \times .746}{.909 \times .87} = 806$$

SIM 4-2

As an initial approximation for sizing a transformer assume that a horsepower equals a KVA. Compare answer with SIM 4-1.

Answer

Total HP = 854

HP = KVA = 854

Problem answer = 806

For an initial estimate, equating horsepower to KVA is done in industry. This accuracy is in general good enough since motor horsepowers will probably change before the design is finished. The load at which the motor is operating is not established at the beginning of a project and this approach usually gives a conservative answer.

SIM 4-3

It is desired to operate the plant of SIM 4-1 at a power factor of .95. What approximate capacitor bank is required?

Answer

From SIM 4-1 the plant is operating at a power factor of .87. The power factor of .87 corresponds to an angle of 29°.

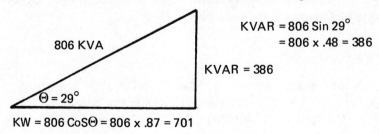

KVAR = 806 Sin 29°

= 806 x .48 = 386

KVAR = 386

806 KVA

Θ = 29°

KW = 806 CosΘ = 806 x .87 = 701

A power factor of .95 is required.

$$Cos\Theta = .95$$
$$\Theta = 18°$$

The KVAR of 386 needs to be reduced by adding capacitors.

Remember KW does not change with different power factors, but KVA does.

Thus, the desired power triangle would look as follows:

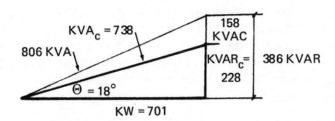

$$CoS\Theta = .95$$
$$\Theta = 18° \qquad Sin\ 18° = .31$$

$$KVA_c = \frac{701}{CoS\Theta} = \frac{701}{.95} = 738$$

Note: Power factor correction reduces total KVA

$$KVAR_c = 738\ Sin\ 18° = 738 \times .31 = 228$$

Capacitance Bank = 386 - 228
$$\cong 158\ KVAC$$

SIM 4-4

Assume that all motors in SIM 4-1 are not running at the same time. The diversity factor which takes into account the cycle time is assumed from previous plant experience to be 1.1. Indicate the minimum transformer size.

Answer

$$KVA_{Min.} = \frac{KVA_+}{Diversity\ Factor} = \frac{806}{1.1} = 732$$

Many times in industry the transformer capacity is simply based on the sum of the motor horsepowers plus an additional factor to take into account growth. The conservative sizing approach may not be too exact, but it does allow for normal changes in design and growth capacity. Remember electrical loads seldom shrink.

WHERE TO LOCATE CAPACITORS

As indicated, the primary purpose of capacitors is to reduce the reactive power levels. Additional benefits are derived by capacitor location. Figure 4-1 indicates typical capacitor locations. Maximum benefit of capacitors is derived by locating them as close as possible to the load. At this location, its kilovars are confined to the smallest possible segment, decreasing the load current. This, in turn, will reduce power losses of the system substantially. Power losses are proportional to the square of the current. When power losses are reduced, voltage at the motor increases; thus, motor performance also increases.

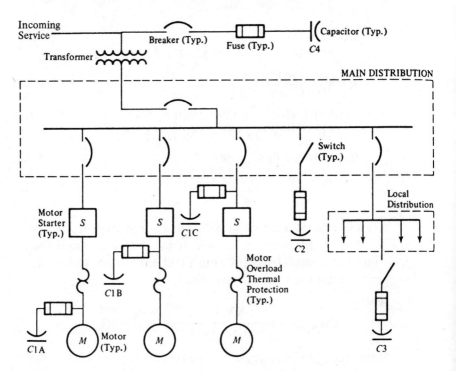

Figure 4-1. Power Distribution Diagram Illustrating Capacitor Locations

Locations $C1A$, $C1B$ and $C1C$ of Figure 4-1 indicate three different arrangements at the load. Note that in all three locations, extra switches are not required, since the capacitor is either switched with the motor starter or the breaker before the starter. Case $C1A$ is recommended for new installations, since the maximum benefit is derived and the size of the motor thermal protector is reduced. In Case $C1B$, as in Case $C1A$, the capacitor is energized only when the motor is in operation. Case $C1B$ is recommended in cases where the installation is existing since the thermal protector does not need to be re-sized. In position $C1C$, the capacitor is permanently connected to the circuit, but does not require a separate switch, since it can be disconnected by the breaker before the starter.

It should be noted that the rating of the capacitor should *not* be greater than the no-load magnetizing KVAR of the motor. If this condition exists, damaging overvoltage or transient torques can occur. This is why most motor manufacturers specify maximum capacitor ratings to be applied to specific motors.

The next preference for capacitor locations as illustrated by Figure 4-2 is at locations $C2$ and $C3$. In these locations, a breaker or switch will be required. Location $C4$ requires a high voltage breaker. The advantage of locating capacitors at power centers or feeders is that they can be grouped together. When several motors are running intermittently, the capacitors are permitted to be on line all the time, reducing the total reactive power regardless of load.

POWER FLOW CONCEPT

Power flowing is analogous to water flowing in a pipe. To supply several small water users, a large pipe services the plant at a high pressure. Several branches from the main pipe service various loads. Pressure reducing stations lower the main pressure to meet the requirements of each user. Similarly, a large feeder at a high voltage services a plant. Through switchgear breakers, the main feeder is distributed into smaller feeders. The switchgear breakers serve as a protector for each of the smaller feeders. Transformers are used to lower the voltage to the nominal value needed by the user.

HOW TO DRAW A ONE-LINE DIAGRAM

An overall one-line diagram indicates where loads are located and how they are fed.

• The first step is to establish loads and their locations by communicating with the various engineers. (Remember initial design is based on your best estimate.)

• The next step is to determine the incoming voltage level based on available voltages from the utility company and the distribution voltage within the plant.

(a) For single buildings and small complexes without heavy equipment loads, incoming voltage levels may be 208 or 480 volts. For small industrial plants to 10,000 KVA, voltage levels may be 2300, 4160, 6900 or 13.8 KV.

(b) For medium plants 10,000 KVA to 20,000 KVA voltage levels may be 13.8 KV.

(c) For large plants above 20,000 KVA, 13.8 KV or 33 KV are typical values.

The advantages with the higher voltage levels are:

(a) Feeders and feeder breakers can handle greater loads. (More economical at certain loads.)

(b) For feeders which service distant loads, voltage drops are not as noticeable on the higher voltage system.

• The third step is to establish equipment types, sizes and ratings.

• The last step is to determine the system reliability required. The type of process and plant requirements are the deciding factors. The number of feeds and the number of transformers determine the degree of reliability of a system.

The three commonly used primary distribution systems for industrial plants are the simple radial, primary selective, and secondary selective systems.

• *The Simple Radial System* is the most economical. As Figure 4-2 indicates, it is comprised of one feed and one transformer.

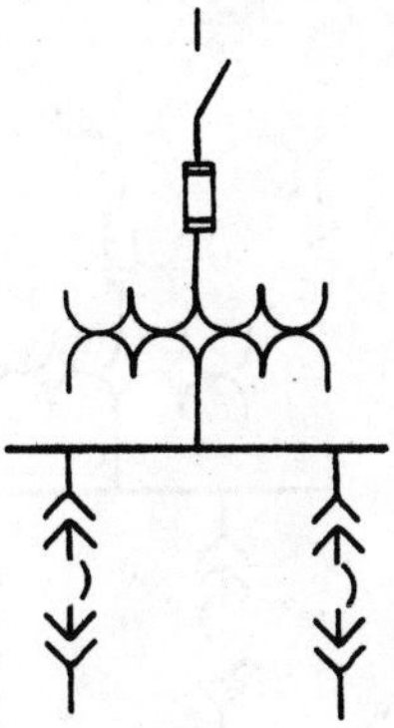

Figure 4-2
Simple Radial System

• *The Primary Selective System* is comprised of two feeds and two primary transformer disconnect switches. See Figure 4-3.

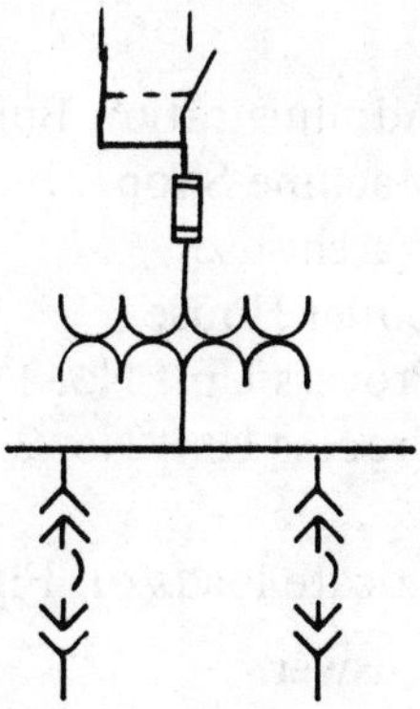

Figure 4-3
Primary Selective System

• *The Secondary Selective System* is the most reliable and the most expensive. As Figure 4-4 indicates, it is comprised of two complete substations joined by a tie breaker.

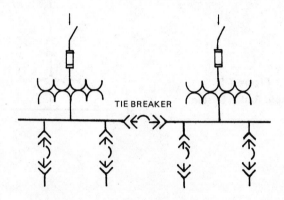

Figure 4-4. Secondary Selective System

SIM 4-5

Develop the one-line diagram for the Ajax Plant. The various steps for developing a one-line diagram are illustrated in SIM 4-5 through SIM 4-9.

Step 1: Establish loads and their locations.

	Approximate Load (KVA)
Administration Building	210
Machine Shop	340
Warehouse	185
Boiler House	675
Process Unit No. 1	890
Process Unit No. 2	765

Locate loads on Figure 4-5.

Answer

See Figure 4-6.

SIM 4-6

Step 2: Determine distribution voltage to substations for SIM 4-5. Assume plant capacity may triple in the future and the utility primary voltage is 115 KV.

Answer

Total load is 3065 which represents the small plant category. Since the plant load may triple in the future, **13.8 KV** would be recommended since it meets the present load with future expansion.

SIM 4-7

Step 3: Establish equipment sizes for SIM 4-5. Substations should be located as close as possible to the load center. If loads are small (below 300 KVA) and are located near another load center, consideration should be given to combining it with other small loads.

Use standard size transformers illustrated in Table, and size at least 1.25 times given load.

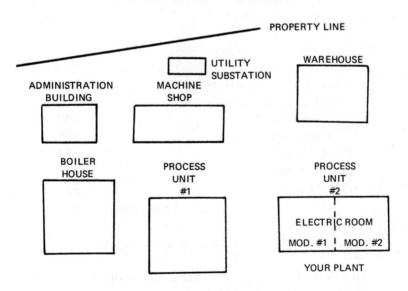

Figure 4-5. Ajax Plant Layout

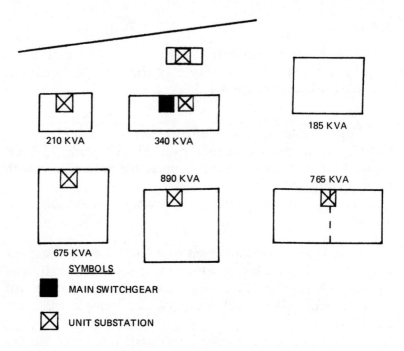

Figure 4-6. Answers to SIM 4-5

	Load
Administration Building	210
Feed from same Transformer	
Machine Shop	340
Warehouse	185
Boiler House	675
Process Unit No. 1	890
Process Unit No. 2	_765_
Total Load	3065

Locate substations and switchgear on Figure 4-5.

Step 4: Comment on the system reliability recommended assuming a noncritical process.

Answer

	Load 1.25	Transformer Size
Administration Building	262	300 KVA
Feed from same Transformer		
Machine Shop	425 ⎫	
Warehouse	230 ⎭	750 KVA
Boiler House	845	1000 KVA
Process Unit No. 1	1112	1500 KVA
Process Unit No. 2	956	1000 KVA

The main feeder from the utility transformer should be sized to meet the total load of 3065 plus capacity for the future. It becomes impractical to run this size feeder to each substation. Thus the main switchgear is provided to distribute power to the various substations.

The substations and switchgear are located on Figure 4-6.

Unless the process is critical, a simple radial system is commonly used.

SIM 4-8

Draw a one-line diagram for SIM 4-5. Assume that up to 500 KVA can be put on each motor control center. Provide two switchgear breakers, one feeding the process substations and the second feeding the utility and auxiliary areas.

Answer   See Figure 4-7.

68 Introduction To Efficient Electrical Systems Design

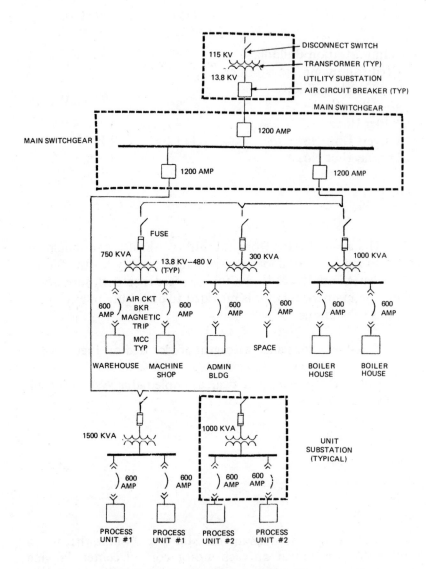

Figure 4-7. Overall One-Line Diagram

GENERAL TIPS FOR DISTRIBUTION SYSTEMS

- Always size unit substations with growth capacity (25% growth capacity for transformers is common practice).
- A transformer with fans increases its rating. A 1000 KVA dry-type transformer with fans is good for 1333 KVA (33% increase). For an oil-type transformer a factor of 25% is used. Fans should only be considered for emergency conditions or for expanding existing plants.
- The question comes up as to where to locate equipment. Incoming switchgear is usually located near the property line so that the utility company can gain easy access to the equipment. Substations and motor control centers are usually located indoors in electrical rooms.

ELECTRICAL ROOMS

The electrical engineer should keep in mind the following when specifying electrical room requirements.
- Do not allow roof penetrations. Any roof opening increases the risk of fluid entering the electrical equipment.
- Do not allow other trades to use electrical room space.
- General ambient temperature should be 40°C. (Special equipment, such as computers, may require air conditioning.)
- Lay out electrical rooms with the following in mind.
  - (a) Sufficient aisle space and door clearances should be provided to allow for maintenance and replacing of damaged transformers and breakers. Use the recommended clearances established by the vendor.
  - (b) Double doors of adequate height (usually 8 feet) should be provided at exits in order to remove equipment.

SHORT-CIRCUIT CURRENTS

Faults occur for many reasons; deterioration of insulation, accidents, rats electrocuted across power leads, equipment failure and a multitude of other events. When a fault occurs a large short-circuit current flows. At first it has an initial peak or asymmetrical value, but after a period of time it will become symmetrical about the zero axis. Equipment must be rated to meet both the full-load currents and the short-circuit currents available.

RATING SUBSTATION BREAKERS (480V)

All breakers should be sized to meet the full-load current, the available short-circuit current, and must be able to coordinate with the system. Coordination of protective devices such as breakers means:

(a) That a protective device will not trip under normal operating conditions such as when the motor is started.

(b) That the protective element closest to the fault will open before the other devices upstream.

The impedance of the transformer limits the amount of short circuit current which could flow. Table 4-1 illustrates how the transformer rating and utility system rating affect the short-circuit current. Use this table as a guide. A more detailed analysis is required on actual selection.

SIM 4-9

For a 1000 KVA transformer, 5.75% impedance, available primary short circuit of 250 MVA, determine the short-circuit current assuming 100% motor contribution.

Answer

From Table 4-1, the short-circuit current is 24,400 symmetrical amperes.

Table 4-1. Short Circuit Application Table: 480 Volts, Three Phase

Transformer Rating 3-Phase KVA and Impedance Percent	Maximum Short Circuit MVA Available from Primary System	Normal Load Continuous Current Amp	Short Circuit Current RMS Symmetrical Amp			Long-Time Instantaneous _Recommended_ Min. Breaker Frame
			Transformer Alone	100% Motor Load	Combined	
1	2	3	4	5	6	7
300 5%	50	361	6500	1400	7900	225
	100		6900		8300	
	150		7000		8400	
	250		7100		8500	
	500		7200		8600	
	750		7200		8600	
	Unlimited		7300		8700	
500 5%	50	601	10000	2400	12400	225
	100		10900		13300	
	150		11300		13700	
	250		11600		14000	
	500		11800		14200	
	750		11800		14200	
	Unlimited		12000		14400	
750 5.75%	50	902	12500	3600	16100	225
	100		13900		17500	
	150		14400		18000	
	250		14900		18500	
	500		15300		18900	
	750		15400		19000	
	Unlimited		15700		19300	

|more|

kVA / %Z						
1000 5.75%	50	1203	15500	4800	20300	225
	100		17800		22600	600
	150		18800		23600	
	250		19600		24400	
	500		20200		25000	
	750		20500		25300	
	Unlimited		20900		25700	
1500 5.75%	50	1804	20600	7200	27800	600
	100		24900		32100	1600
	150		26700		33900	
	250		28400		35600	
	500		29800		37000	
	750		30300		37500	
	Unlimited		31400		38600	
2000 5.75%	50	2406	24700	9600	34300	1600
	100		31100		40700	
	150		34000		43600	
	250		36700		46300	
	500		39100		48700	
	750		40000		49600	3000
	Unlimited		41900		51500	
2500 5.75%	50	3008	28000	12000	40000	1600
	100		36400		48400	
	150		40500		52500	3000
	250		44500		56500	
	500		48100		60100	
	750		49500		61500	
	Unlimited		52300		64300	
3000 5.75%	50	3607	30700	14400	45100	1600
	100		41200		55600	3000
	150		46500		60900	
	250		51900		66300	
	500		56800		71200	4000
	750		58700		73100	
	Unlimited		62700		77100	

Table 4-1. Short Circuit Application Table: 480 Volts, Three Phase (concluded)

Breaker Frame	Continuous Current Ratings Typical Trip Sizes	480V Breaker Rating	Short-Time Rating Amperes RMS Symmetrical
8	9	10	11
225	15, 20, 30, 40, 50, 70, 90, 100, 125, 150, 175, 200, 225	225	9,000
600	40, 50, 70, 90, 100, 125, 150, 175, 200, 225, 250, 300, 350, 400, 500, 600	600	22,000
1600	200, 225, 250, 275, 300, 350, 400, 500, 800, 1000, 1200, 1600	1600	50,000
3000	2000, 2500, 3000	3000	65,000
4000	2000, 2500, 3000, 4000	4000	85,000

Note: Usually at the beginning of a project if no utility data is available, assume unlimited short-circuit current and 100% motor load contribution.

SIM 4-10

A 1500 KVA substation secondary breaker feeds a 600 ampere Motor Control Center bus. Select a breaker frame to meet this load.

Answer

From Table 4-1, the 1600 ampere breaker is the minimum breaker size.

Based on a continuous current rating a 600 ampere frame breaker would have been sufficient, but the 600 ampere breaker can only handle a short-circuit current of 22,000 amperes. The 1600 ampere breaker can handle a short-circuit current up to 50,000 amperes and is good for 1600 continuous amperes. Thus, the frame size required is 1600 amperes. The trip rating which indicates when the breaker will open can be set at any value as indicated in Table 4-1. To protect the motor control center, a 600 ampere trip would be chosen.

ANALYZING BIDS

It is the responsibility of the electrical engineer to recommend vendors to build the electrical equipment for the plant. To get competitive bids, requests for quotations are initiated with several bidders. After the bids have been received, they must be evaluated in terms of quality, costs and schedule in particular.

• Check to see if vendor has met specifications. Check capacities, rating, etc.

- Look at advantages and disadvantages with each vendor. Physical size, energy consumption, noise level, weight, etc. may affect the final recommendation.

- Analyze delivery schedule.

- Evaluate cost picture. Check if all items are included.

Sometimes a client may wish to direct purchase an item due to overriding factors, such as spare parts or to match an existing installation.

JOB SIMULATION – SUMMARY PROBLEM

Job 3

(a) The client wishes to know the power factor at which the plant is operating. Exclude motors below 3 HP from computations. Assume a lighting load of 40 KW. The plant is comprised of two identical modules (2 motors for each equipment number listed in Job 1). (Remember that KVAs at different power factors can not be added directly.)

(b) Based on the total KVA of the plant determine the transformer size and the rating of each breaker. Assume an individual breaker feeds each module. The substation data will be sent to the three industries listed below for competitive bids:

Recommended vendors: ABC Industries
DEF Industries
GHI Industries

(c) Based on the substation quotes received in Job 3b, recommend a vendor. Required delivery date is 35 weeks from date of order.

Complete.

Simplified Bid Analysis Form

	Vendor's Name	Vendor's Name	Vendor's Name
Cost per Substation			
Delivery			
Meets Specification			
Overall Area			

Bidder
Recommended

Reason:

(a)   Based on the motor list, the plant power factor was estimated at .87. Here's what the client expected to see:

Module #1

Lighting $KW_3 = 40$ Total

Motors 3-30	Motors 40-100
30	60
10	100
25	40
20	50
20	250
20	
5	
130	

At Full Load:

P.F. = 83.5	P.F. = 87.4
η = 86.2	η = 90.9

$$KVA_1 = \frac{130 \times .746}{.83 \times .86} = 135 \qquad KVA_2 = \frac{250 \times .746}{.90 \times .87} = 238$$

$$
\begin{aligned}
KW_1 &= KVA \ CoS\Theta \\
&= KVA \ .83 \\
&= 112
\end{aligned}
\qquad
\begin{aligned}
KW_2 &= KVA_2 \ CoS\Theta \\
&= KVA \ .87 \\
&= 207
\end{aligned}
$$

$$KVAR_1 = KVA_1 Sin\Theta \qquad KVAR_2 = KVA_2 Sin\Theta$$

$$\Theta = 33° \qquad\qquad\qquad \Theta = 29°$$

$$
\begin{aligned}
KVAR_1 &= KVA_1 \times .54 \\
&= 135 \times .54 \\
&= 73
\end{aligned}
\qquad
\begin{aligned}
KVAR_2 &= KVA_2 \times .48 \\
KVAR_2 &= 115
\end{aligned}
$$

$$KW_{total} = KW_1 + KW_2 + KW_1 + KW_2 + KW_3 =$$

$$\qquad\qquad \underset{1}{\text{Module}} \qquad \underset{2}{\text{Module}} \qquad\qquad = 678 \ KW$$

$$KVAR_{total} = KVAR_1 + KVAR_2 + KVAR_1 + KVAR_2$$
$$\quad\quad\quad\quad \text{Module 1} \quad\quad\quad \text{Module 2}$$
$$= 73 + 115 + 73 + 115 = 376$$

$$KVA_{total} = \sqrt{(678)^2 + (376)^2} = 774 \text{ KVA}$$

$$Cos\Theta = \frac{KW_{total}}{KVA_{total}} = \frac{678}{774} = .87$$

(b) The substation requisition should have included a 1000 KVA transformer and two 600 ampere feeder breakers.

Transformer = 1.25 (Total KVA) = 1.25 X 774
= 967 KVA size required.

The closest transformer size is 1000 KVA.

The minimum breaker size based on Table 4-1 is 600 amperes. This size is required even though the total load on each breaker is $\frac{774}{2}$ = 387 KVA or

$$\frac{387K}{\sqrt{3} \text{ X } 480} = 466 \text{ Amps}$$

The breaker must be sized to meet the short-circuit current of approximately 25,000 amps symmetrical.

Quotes received based on the substation requisition:

ABC Industries

Dear Sir:

In reply to subject inquiry, we are pleased to quote on the following:

1–480 Volt Substation including:

1. Incoming line compartment with load interrupter switch and current limiting fuse.

2. Dry type transformer 1000 KVA 13.8 KV to 480 volt wye.

3. Low voltage switchgear with main bus, feeder instrument and metering—2 breakers.

Equipment shall be arranged in accordance with layout attached.

Total Price is $37,000.

Shipment can be made 30 weeks after receipt of order.

Standard terms 30 days from date of invoice.

Very truly yours,

/s/ Al B. See

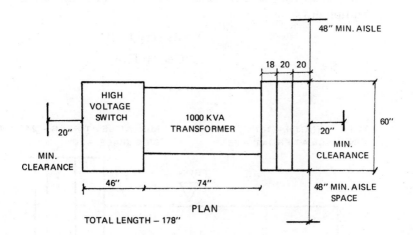

DEF Industries

Gentlemen:

We regret to inform you that at this time we can not quote. We appreciate your interest in DEF Industries and hope that we can be of service in the future.

Very truly yours,

/s/ Dee Frank

GHI Industries

Dear Sir:

We wish to offer our package completely in accordance with your specification.

The total price for the order is $38,500.

The above prices are quoted F.O.B. factory.

The earliest time equipment can be shipped is 40 weeks after receipt of order.

The attached layout shows the arrangement of the equipment.

Yours very truly,

/s/ Gee Hi Eye

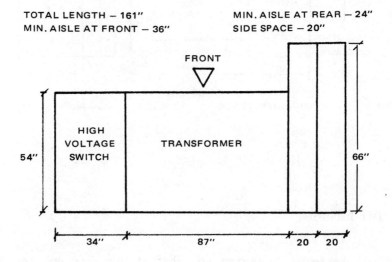

TOTAL LENGTH — 161″   MIN. AISLE AT REAR — 24″
MIN. AISLE AT FRONT — 36″   SIDE SPACE — 20″

FRONT

HIGH VOLTAGE SWITCH   TRANSFORMER

54″   66″

34″   87″   20   20

JOB 4c: A Completed Bid Analysis Form

	Vendor's Name	Vendor's Name	Vendor's Name
	ABC Ind.	DEF Ind.	GHI Ind.
Cost per Substation	$37,000	Declined	$38,500
Delivery	30 weeks		40 weeks
Meets Specification	Yes		Yes
Overall Area	178" X 60"		161" X 66"

Bidder
Recommended   ABC   Industries

Reason: Only vendor which meets required delivery.
              Lowest price.

SUMMARY

Transformer ratings are based on KVA. Since loads constantly change, a sizing based on the sum of the motor and miscellaneous loads plus 1.25% gives a reasonable initial basis for determining capacity.

When determining the rating of equipment, always make sure that the equipment meets the short-circuit current of the plant.

The one-line diagram is the single most important drawing for the electrical design. It serves as the basis for:

Short-circuit calculations

Coordination studies

Power factor correction

Equipment selection

Power plans

In addition, the one-line diagram is a must for plant maintenance. By use of the one-line diagram, problem areas can be located and corrective action can be taken.

Chapter 5

CONDUIT AND
CONDUCTOR SIZING

CONDUCTORS FOR USE IN RACEWAYS

Once the facility environment has been established, cable insulation can be selected. Table 5-1 lists a number of insulations commonly used. Each insulation has a specified operating temperature. Types THW and RHW are frequently used in industrial plants for 600 volt or less distribution.

Allowable ampacities of copper conductors in conduit at 30°C are shown in Table 5-2. Sometimes the designations 0, 00, 0000 are referred to as 1/0, 2/0, 3/0, 4/0, respectively.

Tables 5-2 through 5-6 are reproduced by permission from the National Electrical Code 1984, copyright 1983 National Fire Protection Association. Copies of the complete code are available from the Association, 470 Lexington Avenue, Boston, Massachusetts 02110.

SIM 5-1

For (3) 350 MCM THW copper conductors, run in conduit at 30°C, determine the conductor capacity.

Answer

From Table 5-2 type THW is rated for 310 amperes.

Table 5-1. Conductors for Use in Raceways
(A Partial Listing)

TYPE	INSULATION
FEP	Fluorinated Ethylene Propylene
MI	Mineral Insulation (metal sheathed)
RH	Heat-Resistant Rubber
RHH	Heat-Resistant Rubber
RHW	Moisture and Heat-Resistant Rubber
RUH	Heat-Resistant Latex Rubber
RUW	Moisture-Resistant Latex Rubber
T	Thermoplastic
TW	Moisture-Resistant Thermoplastic
THW	Moisture and Heat-Resistant Thermoplastic
THWN	Moisture and Heat-Resistant Thermoplastic
THHN	Moisture and Heat-Resistant Thermoplastic
XHHW	Moisture and Heat Resistant Cross-Linked Synthetic Polymer
V	Varnished Cambric
AVA	Asbestos and Varnished Cambric
AVL	Asbestos and Varnished Cambric
AVB	Asbestos and Varnished Cambric

SIM 5-2

What type of insulation is type RHW?

Answer

From Table 5-1 RHW is composed of moisture and heat resistant rubber.

SIM 5-3

Determine the ampacity for (3) 4/0 cross-linked polyethylene copper conductors run in conduit at 86°F in a dry location.

Answer

In Table 5-1 cross-linked polyethylene copper cable is Type XHHW. When it is used in a dry location it is good for 90°C. From Table 5-2, 0000 is good for 260 amperes.

Table 5-2.

Reprinted with permission from NFPA 70-1984, National Electrical Code®, Copyright© 1983, National Fire Protection Association, Quincy, Massachusetts 02269. This reprinted material is not the complete and official position of the NFPA on the referenced subject, which is represented only by the standard in its entirety. National Electrical Code® and NEC® are Registered Trademarks of the National Fire Protection Association, Inc., Quincy, MA.

Table 310-16. Ampacities of Insulated Conductors Rated 0-2000 Volts, 60° to 90°C

Not More Than Three Conductors in Raceway or Cable or Earth (Directly Buried), Based on Ambient Temperature of 30°C (86°F)

Size	Temperature Rating of Conductor, See Table 310-13								Size
	60°C (140°F)	75°C (167°F)	85°C (185°F)	90°C (194°F)	60°C (140°F)	75°C (167°F)	85°C (185°F)	90°C (194°F)	
AWG MCM	TYPES †RUW, †T, †TW, †UF	TYPES †FEPW, †RH, †RHW, †RUH, †THW, †THWN, †XHHW, †USE, †ZW	TYPES V, MI	TYPES TA, TBS, SA, AVB, SIS, †FEP, †FEPB, †RHH †THHN, †XHHW*	TYPES †RUW, †T, †TW, †UF	TYPES †RH, †RHW, †RUH, †THW †THWN, †XHHW, †USE	TYPES V, MI	TYPES TA, TBS, SA, AVB, SIS, †RHH, †THHN, †XHHW*	AWG MCM
	COPPER				ALUMINUM OR COPPER-CLAD ALUMINUM				
18	....	....	....	14	....	....	....	....	....
16	....	....	18	18	....	....	....	....	....
14	20†	20†	25	25†	....	....	....	....	
12	25†	25†	30	30†	20†	20†	25	25†	12
10	30	35†	40	40†	25	30†	30	35†	10
8	40	50	55	55	30	40	40	45	8
6	55	65	70	75	40	50	55	60	6
4	70	85	95	95	55	65	75	75	4
3	85	100	110	110	65	75	85	85	3
2	95	115	125	130	75	90	100	100	2
1	110	130	145	150	85	100	110	115	1
0	125	150	165	170	100	120	130	135	0
00	145	175	190	195	115	135	145	150	00
000	165	200	215	225	130	155	170	175	000
0000	195	230	250	260	150	180	195	205	0000
250	215	255	275	290	170	205	220	230	250
300	240	285	310	320	190	230	250	255	300
350	260	310	340	350	210	250	270	280	350
400	280	335	365	380	225	270	295	305	400
500	320	380	415	430	260	310	335	350	500
600	355	420	460	475	285	340	370	385	600
700	385	460	500	520	310	375	405	420	700
750	400	475	515	535	320	385	420	435	750
800	410	490	535	555	330	395	430	450	800
900	435	520	565	585	355	425	465	480	900
1000	455	545	590	615	375	445	485	500	1000
1250	495	590	640	665	405	485	525	545	1250
1500	520	625	680	705	435	520	565	585	1500
1750	545	650	705	735	455	545	595	615	1750
2000	560	665	725	750	470	560	610	630	2000
	AMPACITY CORRECTION FACTORS								
Ambient Temp. °C	For ambient temperatures other than 30°C, multiply the ampacities shown above by the appropriate factor shown below.								Ambient Temp. °F
31-40	.82	.88	.90	.91	.82	.88	.90	.91	87-104
41-45	.71	.82	.85	.87	.71	.82	.85	.87	105-113
46-50	.58	.75	.80	.82	.58	.75	.80	.82	114-122
51-60	....	.58	.67	.71	....	.58	.67	.71	123-141
61-70	....	.35	.52	.58	....	.35	.52	.58	142-158
71-80	....	....	.30	.41	....	....	.30	.41	159-176

† The overcurrent protection for conductor types marked with an obelisk (†) shall not exceed 15 amperes for 14 AWG, 20 amperes for 12 AWG, and 30 amperes for 10 AWG copper; or 15 amperes for 12 AWG and 25 amperes for 10 AWG aluminum and copper-clad aluminum after any correction factors for ambient temperature and number of conductors have been applied.

* For dry locations only. See 75°C column for wet locations.

CONDUCTOR DERATINGS

Derating Due To The Number of Conductors

If three conductors or less are run in a conduit, no derating need be applied. Based on the number of conductors above three, additional deratings must be used. For example: 4 to 6 conductors derate the cable 80%; 7 to 24, 70%; 25 to 42, 60%; and 43 and above, 50%. Because of these deratings and ease in pulling, many times six conductors servicing a load are run in two conduits.

Derating Due To Ambient Temperature

If the ambient differs from 30°C, another derating should be applied. These correction factors are included in Table 5-2.

SIM 5-4

What is the allowable ampacity of (6) 500 MCM, THW, copper conductors run in the same conduit at 50°C?

Answer

From Table 5-2, the allowable ampacity of 500 MCM cable is 380 amperes.

Derating factor for six conductors is .8.

Allowable ampacity = (380 X .8 X .75) = 228 amps

Observation

It should be noted from Table 5-2 that some conductor sizes are not practical. Doubling the area of a conductor, i.e., 500 MCM cable to a 1000 MCM cable, does not double the ampacity rating. In fact doubling the area only results in 50% more capacity. Using conductors larger than 500 MCM is also not recommended due to difficulties in installation. Consequently, if more capacity is required than can be provided by a single 500 MCM conductor, multiple conductors are used in parallel.

CABLE SIZING

Sizing a Feeder to a Motor for Continuous Service

The minimum size cable for power conductors is #12. The cable capacity for a motor should be equal to 1.25 X the full load amperes of the motor. Typical full-load currents are illustrated in Chapter 3, Table 3-2.

SIM 5-5

What cable size should be used for a 50 HP induction motor? Assume the cable is Type THW and run in conduit. Ambient temperature 30°C.

Answer

From Table 3-2, Chapter 3, a 50 HP motor has a F.L.A. (full load amperes) of 65 amps. Cable must be equal to or above 1.25 X 65 = 81 amp. Use #4 AWG.

Sizing Feeder to Several Motors

The size of the feeder which has more than one motor on it is based on 1.25 times the full-load current of the largest motor plus the full-load current of the others.

SIM 5-6

A feeder supplies a 25 HP motor and two 30 HP motors. Determine the size of the feeder if it is rated for ambient 30°C and cable run in conduit.

Answer

F.L.A. – 30 HP motor – 40 amp.
F.L.A. – 25 HP motor – 34 amp.
Cable must meet: 1.25 X 40 + 34 + 40 = 124 amps
The nearest cable size from Table 5-2 is #1 THW.

SIZING A CONDUIT

Once the conductors are selected, they must be protected from physical damage through the use of an enclosed conduit. Depending on the application the material used for the conduit can be rigid steel, electrical metallic tubing (EMT), flexible metal or PVC. Note that for metallic materials, the conduit often serves as the equipment ground.

The size of a conduit depends on the allowable percent fill of the conduit area. Table 5-3 illustrates allowable fills. Table 5-4 is based on Table 5-3 and enables the engineer to readily select the number of conductors which can be installed in a conduit.

Table 5-3.

NEC Table 1.   Percent of Cross Section of Conduit and Tubing for Conductors

(See Table 2 for Fixture Wires)

Number of Conductors	1	2	3	4	Over 4
All conductor types except lead-covered (new or rewiring)	53	31	40	40	40
Lead-covered conductors	55	30	40	38	35

Reprinted with permission from NFPA 70-1984, National Electrical Code®, Copyright© 1983, National Fire Protection Association, Quincy, Massachusetts 02269. This reprinted material is not the complete and official position of the NFPA on the referenced subject, which is represented only by the standard in its entirety. National Electrical Code® and NEC® are Registered Trademarks of the National Fire Protection Association, Inc., Quincy, MA.

Table 5-4 stipulates how many of the same size conductors can be run in a conduit based on the allowable fills mentioned in Table 5-3.

SIM 5-7

What is the allowable percent fill for three conductors Type THW run in conduit.

Answer   From Table 5-3, 40%.

SIM 5-8

A control cable consisting of thirty-seven #14 wires Type TW is run from M.C.C. No. 1 to Process Panel No. 2. What size conduit is required?

Answer

From Table 5-3 the maximum conduit fill is 40%. From Table 5-4 the answer can be read directly. Since 40 wires can be run in a 1¼″ conduit, the correct size is 1¼″.

Table 5-4.

Reprinted with permission from NFPA 70-1984, National Electrical Code® Copyright© 1983, National Fire Protection Association, Quincy, Massachusetts 02269. This reprinted material is not the complete and official position of the NFPA on the referenced subject, which is represented only by the standard in its entirety.

Table 3A. Maximum Number of Conductors in Trade Sizes of Conduit or Tubing (Based on Table 1, Chapter 9)

Type Letters	Conductor Size AWG, MCM	½	¾	1	1¼	1½	2	2½	3	3½	4	5	6
TW, T, RUH, RUW, XHHW (14 thru 8)	14	9	15	25	44	60	99	142	171				
	12	7	12	19	35	47	78	111	131	176			
	10	5	9	15	26	36	60	85					
	8	2	4	7	12	17	28	40	62	84	108		
RHW and RHH (without outer covering), THW	14	6	10	16	29	40	65	93	143	192			
	12	4	8	13	24	32	53	76	117	157			
	10	4	6	11	19	26	43	61	95	127	163		
	8	1	3	5	10	13	22	32	49	66	85	133	
TW, T, THW, RUH (6 thru 2), RUW (6 thru 2)	6	1	2	4	7	10	16	23	36	48	62	97	141
	4		1	3	5	7	12	17	27	36	47	73	106
	3		1	2	4	6	10	15	23	31	40	63	91
	2		1	2	4	5	9	13	20	27	34	54	78
	1			1	3	4	6	9	14	19	25	39	57
FEPB (6 thru 2), RHW and RHH (without outer covering)	0				2	3	5	8	12	16	21	33	49
	00				1	3	5	7	10	14	18	29	41
	000				1	2	4	6	9	12	15	24	35
	0000					1	3	5	7	10	13	20	29
	250					1	2	4	6	8	10	16	23
	300						2	3	5	7	9	14	20
	350						1	3	4	6	8	12	18
	400						1	2	4	5	7	11	16
	500							1	3	4	6	9	14
	600							1	3	4	5	7	11
	700								2	3	4	7	10
	750								2	3	4	6	9

Table 5-4. (concluded)

Table 3B.   Maximum Number of Conductors in Trade Sizes of Conduit or Tubing
(Based on Table 1, Chapter 9)

Type Letters	Conductor Size AWG, MCM	½	¾	1	1¼	1½	2	2½	3	3½	4	5	6
THWN,	14	13	24	39	69	94	154						
	12	10	18	29	51	70	114	164					
	10	6	11	18	32	44	73	104	160				
	8	3	5	9	16	22	36	51	79	106	136		
THHN, FEP (14 thru 2), FEPB (14 thru 8), PFA (14 thru 4/0), PFAH (14 thru 4/0), Z (14 thru 4/0), XHHW (4 thru 500MCM)	6	1	4	6	11	15	26	37	57	76	98	154	
	4	1	2	4	7	9	16	22	35	47	60	94	137
	3	1	1	3	6	8	13	19	29	39	51	80	116
	2		1	3	5	7	11	16	25	33	43	67	97
	1			1	3	5	8	12	18	25	32	50	72
	0			1	3	4	7	10	15	21	27	42	61
	00			1	2	3	6	8	13	17	22	35	51
	000			1	1	3	5	7	11	14	18	29	42
	0000					2	4	6	9	12	15	24	35
	250				1	1	3	4	7	10	12	20	28
	300				1	1	3	4	6	8	11	17	24
	350					1	2	3	5	7	9	15	21
	400						1	3	5	6	8	13	19
	500						1	2	4	5	7	11	16
	600						1	1	3	4	5	9	13
	700						1	1	3	4	5	8	11
	750						1	1	2	3	4	7	11
XHHW	6	1	3	5	9	13	21	30	47	63	81	128	185
	600				1	1	1	1	3	4	5	9	13
	700						1	1	3	4	5	7	11
	750						1	1	2	3	4	7	10

Table 5-5.

Table 5. Dimensions of Rubber-Covered and Thermoplastic-Covered Conductors

Size AWG MCM	Types RFH-2, RH, RHH,*** RHW,*** SF-2		Types TF, T, THW,† TW, RUH,** RUW**		Types TFN, THHN, THWN		Types**** FEP, FEPB, FEPW, TFE, PF, PFA, PFAH, PGF, PTF, Z, ZF, ZFF		Type XHHW, ZW††		Types KF-1, KF-2, KFF-1, KFF-2	
	Approx. Diam. Inches	Approx. Area Sq. In.	Approx. Diam. Inches	Approx. Area Sq. In.	Approx. Diam. Inches	Approx. Area Sq. In.	Approx. Diam. Inches	Approx. Area Sq. Inches	Approx. Diam. Inches	Approx. Area Sq. In.	Approx. Diam. Sq. In.	Approx. Area Sq. In.
Col. 1	Col. 2	Col. 3	Col. 4	Col. 5	Col. 6	Col. 7	Col. 8	Col. 9	Col. 10	Col. 11	Col. 12	Col. 13
18	.146	.0167	.106	.0088	.089	.0062	.081	.0052	…	…	.065	.0033
16	.158	.0196	.118	.0109	.100	.0079	.092	.0066	…	…	.070	.0038
14	30 mils .171	.0230	.131	.0135	.105	.0087	.105	.0087	.129	.0131	.083	.0054
14	45 mils .204*	.0327*	.162†	.0206†	…	…	.105	.0087	…	…	…	…
12	30 mils .188	.0278	.148	.0172	.122	.0117	.121	.0115	.146	.0167	.102	.0082
12	45 mils .221*	.0384*	.179†	.0252†	…	…	.121	.0115	…	…	…	…
12	…	…	.168	.0222	…	…	…	…	…	…	…	…
10	.242	.0460	.199†	.0311†	.153	.0184	.142	.0158	.166	.0216	.124	.0121
10	…	…	.245	.0471	…	…	…	…	…	…	…	…
8	.328	.0845	.276†	.0598†	.218	.0373	.186	.0272	.241	.0456	…	…
8	…	…	…	…	…	…	.206	.0333	…	…	…	…
6	.397	.1238	.323	.0819	.257	.0519	.244	.0468	.282	.0625	…	…
4	.452	.1605	.372	.1087	.328	.0845	.292	.0670	.328	.0845	…	…
3	.481	.1817	.401	.1263	.356	.0995	.320	.0804	.356	.0995	…	…
2	.513	.2067	.433	.1473	.388	.1182	.352	.0973	.388	.1182	…	…
1	.588	.2715	.508	.2027	.450	.1590	.420	.1385	.450	.1590	…	…
0	.629	.3107	.549	.2367	.491	.1893	.462	.1676	.491	.1893	…	…
00	.675	.3578	.595	.2781	.537	.2265	.498	.1948	.537	.2265	…	…
000	.727	.4151	.647	.3288	.588	.2715	.560	.2463	.588	.2715	…	…
0000	.785	.4840	.705	.3904	.646	.3278	.618	.3000	.646	.3278	…	…

/more/

Table 5-5. (concluded)

Table 5 (Continued)

Size AWG MCM	Types RFH-2, RH, RHH,*** RHW,*** SF-2		Types TF, T, THW,† TW, RUH,** RUW**		Types TFN, THHN, THWN		Types**** FEP, FEPB, FEPW, TFE, PF, PFA, PFAH, PGF, PTF, Z, ZF, ZFF		Type XHHW, ZW††	
	Approx. Diam. Inches	Approx. Area Sq. In.	Approx. Diam. Inches	Approx. Area Sq. In.	Approx. Diam. Inches	Approx. Area Sq. In.	Approx. Diam. Inches	Approx. Area Sq. Inches	Approx. Diam. Inches	Approx. Area Sq. In.
Col. 1	Col. 2	Col. 3	Col. 4	Col. 5	Col. 6	Col. 7	Col. 8	Col. 9	Col. 10	Col. 11
250	.868	.5917	.788	.4877	.716	.4026	....	....	.716	.4026
300	.933	.6837	.843	.5581	.771	.4669	....	....	.771	.4669
350	.985	.7620	.895	.6291	.822	.5307	....	....	.822	.5307
400	1.032	.8365	.942	.6969	.869	.5931	....	....	.869	.5931
500	1.119	.9834	1.029	.8316	.955	.7163	....	....	.955	.7163
600	1.233	1.1940	1.143	1.0261	1.058	.8791	....	....	1.073	.9043
700	1.304	1.3355	1.214	1.1575	1.129	1.0011	....	....	1.145	1.0297
750	1.339	1.4082	1.249	1.2252	1.163	1.0623	....	....	1.180	1.0936
800	1.372	1.4784	1.282	1.2908	1.196	1.1234	....	....	1.210	1.1499
900	1.435	1.6173	1.345	1.4208	1.259	1.2449	....	....	1.270	1.2668
1000	1.494	1.7530	1.404	1.5482	1.317	1.3623	....	....	1.330	1.3893
1250	1.676	2.2062	1.577	1.9532	....	....	....	....	1.500	1.7671
1500	1.801	2.5475	1.702	2.2751	....	....	....	....	1.620	2.0612
1750	1.916	2.8832	1.817	2.5930	....	....	....	....	1.740	2.3779
2000	2.021	3.2079	1.922	2.9013	....	....	....	....	1.840	2.6590

* The dimensions of Types RHH and RHW.
** No. 14 to No. 2.
† Dimensions of THW in sizes No. 14 to No. 8. No. 6 THW and larger is the same dimension as T.
*** Dimensions of RHH and RHW without outer covering are the same as THW No. 18 to No. 10, solid; No. 8 and larger, stranded.
**** In Columns 8 and 9 the values shown for sizes No. 1 thru 0000 are for TFE and Z only. The right-hand values in Columns 8 and 9 are for FEPB, Z, ZF, and ZFF only.
†† No. 14 to No. 2.

Power and Control Run in The Same Conduit

Frequently the electrical engineer is faced with the problem of sizing a conduit for both power and control wires. An example might be a local STOP-START pushbutton at the motor. The control wires would be run in the same conduit as the power leads if the horsepower is 60 HP or less. Above this size, it becomes impractical to pull the smaller control wires with the larger power conductors. For different size cables, first use Table 5-5 to determine the area of each cable. Then look at Table 5-6 under the appropriate percent fill column and choose a conduit size whose area is equal to or greater than the total conductor area. Note, that control wire insulation must be rated for high voltage if high voltage power wiring is present.

SIM 5-9

Determine the conduit size for the 50 HP motor, SIM 5-5. Assume a local stop-start pushbotton requiring three #14 control cables.

Answer

From SIM 5-5 #4 conductors are required for a 50 HP motor. Table 5-6 shows cable areas for single conductors. Multiply the area by 3 to get the total area. This approximation gives a conservative cable sizing.

$$\text{Power} \quad 3 \times .1087 = .3261$$
$$\text{Control} \quad 3 \times .0206 = \underline{.0618}$$
$$\text{Total} \quad .3879$$

From Table 5-6 based on 40% fill, conduit size is 1¼".

SUMMARY OF DATA

At the beginning of each project a table should be established summarizing all conduit, cable, fuse and switch or breaker sizes for each motor horsepower. At a glance each motor and its associated auxiliaries can be determined. On a large project this type of table saves a considerable amount of time.

Table 5-6.

Reproduced with permission from the National Electrical Code, 1984 edition, copyright 1984, National Fire Protection Association, 470 Lexington Avenue, Boston, MA 02210.

NEC Table 4. Dimensions and Percent Area of Conduit and of Tubing

Areas of Conduit or Tubing for the Combinations of Wires Permitted in Table 1, Chapter 9.

Trade Size	Internal Diameter Inches	Total 100%	Not Lead Covered			Lead Covered				
			2 Cond. 31%	Over 2 Cond. 40%	1 Cond. 53%	1 Cond. 55%	2 Cond. 30%	3 Cond. 40%	4 Cond. 38%	Over 4 Cond. 35%
½	.622	.30	.09	.12	.16	.17	.09	.12	.11	.11
¾	.824	.53	.16	.21	.28	.29	.16	.21	.20	.19
1	1.049	.86	.27	.34	.46	.47	.26	.34	.33	.30
1¼	1.380	1.50	.47	.60	.80	.83	.45	.60	.57	.53
1½	1.610	2.04	.63	.82	1.08	1.12	.61	.82	.78	.71
2	2.067	3.36	1.04	1.34	1.78	1.85	1.01	1.34	1.28	1.18
2½	2.469	4.79	1.48	1.92	2.54	2.63	1.44	1.92	1.82	1.68
3	3.068	7.38	2.29	2.95	3.91	4.06	2.21	2.95	2.80	2.58
3½	3.548	9.90	3.07	3.96	5.25	5.44	2.97	3.96	3.76	3.47
4	4.026	12.72	3.94	5.09	6.74	7.00	3.82	5.09	4.83	4.45
4½	4.506	15.94	4.94	6.38	8.45	8.77	4.78	6.38	6.06	5.56
5	5.047	20.00	6.20	8.00	10.60	11.00	6.00	8.00	7.60	7.00
6	6.065	28.89	8.96	11.56	15.31	15.89	8.67	11.56	10.98	10.11

Area — Square Inches

POWER LAYOUTS

Power layouts are drawn to scale (usually ¼″ = 1′). Conduits are grouped together where possible to form conduit banks. Usually conduits run vertically and horizontally. If conduit and cable sizes cannot be shown on drawing, a separate conduit and cable schedule is required. Conduit layouts should be coordinated with other groups (Piping, HVAC) to avoid interferences.

SIM 5-10

For the plan below, draw a conduit layout.

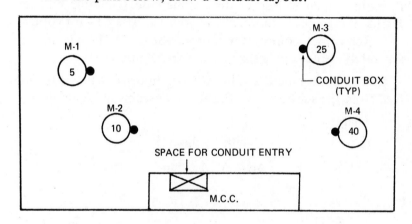

Answer

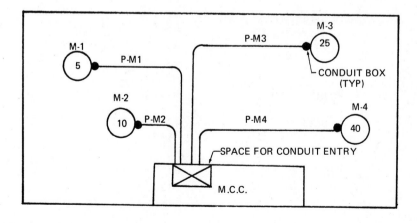

JOB SIMULATION – SUMMARY PROBLEM

Job 5

(a)   With the motor data of Job 1, Chapter 2, establish a conduit and cable schedule. Ambient 40°C and wire Type THW. Assume each stop-start motor requires three #14 for a local pushbutton station. No local control is provided for the two-speed motor, the reversing motors, and AG-1 and FP4. Assume hoist control by others.

When power and control are run in the same conduit, designate the conduit by PC–motor number. For power alone, use P–motor number. For control alone, use C–motor number. Exclude single-phase motors from your list. Use #14 for control.

Remember control for motors above 60 HP will be run separately. Note: Cable size 3 is not frequently used.

(b)   Determine the cable and conduit size required for the feed from the substation to the M.C.C. Assume 40°C ambient.

Analysis

(a)   **Conduit Schedule**

Conduit No.	No.	Size	Conduit Size
P-AG1	1	#2	1¼"
P-CF3	1	# 2/0	1½"
P-FP4	1	#6	1"
PC-TP-5	1 1	#12 } #14 }	¾"
PC-CTP6	1 1	#6 } #14 }	1"
P-CT9	2	#8	¾"
PC-HF10	1 1	#4 } #14 }	¾"
PC-HF11	1 1	#8 } #14 }	1"
PC-BC13	1 1	#2 } #14 }	1¼"
P-C16	1	#8	¾"
P-H17	1	#12	½"
PC-SC19	1 1	#12 } #14 }	¾"

(b) The cable to the M.C.C. should be at least 1.25% of the F.L.A. of the largest motor plus the full load amps of the others. In this case it is approximately 525 amps. The lighting and fractional HP load of approximately 15 amps is then added, bringing the total to 540 amps.

Two 350 MCM run in separate conduits can feed a total load of 2 X 310 X .88 = 545.

For this application, two 2½ conduits are required.

SUMMARY

The National Electrical Code gives specific tables for calculating ampacities and conduit sizes for different cables and various conditions.

In this chapter, conduit and cable determinations have been illustrated, but there are other raceway applications. Wireways and trays are used as a raceway to carry the cable. Appropriate derating factors should be applied, based on the National Electrical Code when using other types of raceways.

Chapter 6

HOW TO DESIGN A LIGHTING SYSTEM

With the increased concern for energy conservation in recent years, much attention has been focused on lighting energy consumption and methods for reducing it. Along with this concern for energy efficient lighting has come the realization that lighting has profound affects on worker productivity as well as important aesthetic qualities. This chapter presents an introduction to lighting design and some of the energy efficient techniques which can be utilized while maintaining the quality of illumination.

LAMP TYPES

There are six different light sources that are popular today: incandescent, fluorescent, mercury vapor, metal halide, high pressure sodium and low pressure sodium. All lamps except incandescent are gas discharge lamps, meaning that light is created through the excitation of gases inside the lamp. All gas discharge lamps require a ballast. A ballast accomplishes the following functions:

1. Limits the current flow.

2. Provides a sufficiently high voltage to start the lamp.

3. Provides the correct voltage to allow the arc discharge to stabilize.

4. Provides power-factor correction to offset partially the the coils' inductive reactance.

Lamp efficacy is determined by the amount of light, measured in lumens, produced for each watt of power the lamp requires. The lumens per watt (LPW) of the various light sources can vary considerably. Table 6-1 shows typical LPW ratings including power consumed by the ballast (ballast losses) where applicable.

Table 6-1. General Lighting Lamp/Ballast Characteristics

Type of Lamp	Wattage Range	Initial Lumens Per Watt Including Ballast Losses	Average Rated Life (Hours)
Low Pressure Sodium	18-180	62-150	12,000-18,000
High Pressure Sodium	35-1,000	51-130	7,500-24,000+
Metal Halide	175-1,500	69-115	7,500-20,000
Mercury Vapor			
Standard	40-1,000	24-60	12,000-24,000+
Self-Ballasted	160-1,250	14-30	10,000-20,000
Fluorescent			
Standard	20-215	63-95	9,000-20,000+
Self-Ballasted	8-44	22-50	7,500-18,000
Incandescent	60-1,500	13-24	750-3,500

INCANDESCENT LAMPS

The incandescent lamp is one of the most common light sources and is also the light source with the lowest efficacy (lumens per watt) and shorest life. This lamp is still popular, however, due to the simplicity with which it can be used and the low price of both the lamp and the fixture. Additionally, the lamp does not require a ballast to condition its power supply, light direction and brightness are easily controlled and it produces light of high color quality.

The most common types of incandescent lamps are: the "A" or standard shaped lamp; the "PS" or pear-shaped lamp; the "R" or refelector lamp; the "PAR" or parabolic-aluminized-reflector lamp and the tungsten-halogen (or quartz) lamp.

Light is produced in an incandescent lamp when the coiled tungsten filament is heated to incandescence (white light) by its resistance to a flow of electric current. The life of the lamp and its light output are determined by its filament temperature. The higher the temperature for a given lamp, the greater the efficacy and the shorter the life. The efficacy of incandescent lamps, however, does increase as the lamp *wattage* increase. This makes it possible to save on both energy and fixture costs whenever you can use one higher wattage lamp instead of two lower wattage lamps.

FLUORESCENT LAMPS

The fluorescent lamp is becoming the most common light source. It is easily distinguished by its tubular design—circular, straight or bent in a "U" shape. In operation, an electric arc is produced between two electrodes which can be several feet apart depending on the length of the tube. The ultraviolet light produced by the arc activates a phosphor coating on the inside wall of the tube, causing light to be produced.

Unlike the incandescent lamp, the fluorescent lamp requires a ballast to strike the electric arc in the tube initially and to maintain that arc. Proper ballast selection is important to optimum light output and lamp life.

Lamp sizes range from four watts to 215 watts. The efficacy (lumens per watt) of a lamp increases with lamp length. Reduced wattage fluorescent lamps and ballasts introduced in the last few years use from 10 percent to 20 percent less wattage than conventional fluorescent lamps.

Fluorescent lamps are available in a wide variety of colors but for most application the cool white, warm white and (newly introduced lite white) lamps produce acceptable color and high efficacy. Since fluorescent lamps are linear light sources with relatively low brightness as compared with point sources (incandescent and high intensity discharge lamps), they are suited for indoor application where lighting quality is important and ceiling heights are moderate.

Fluorescent lamp life is rated according to the number of operating hours per start, for example, 20,000 hours at three hours operation per start. The greater number of hours per start, the greater the lamp life. Because fluorescent lamp life ratings have increased, however, the number of times you turn a lamp on or off has become less important. As a general rule, if a space is to be unoccupied for more than a few minutes, the lamps should be turned off.

HIGH INTENSITY DISCHARGE LAMPS

High intensity discharge (HID) is the term used to designate four distinct types of lamps (mercury vapor, metal halide, high pressure sodium and low pressure sodium). Like fluorescent lamps they produce light by establishing an arc between two electrodes; however, in HID lamps the electrodes are only a few inches apart.

HID lamps require a few minutes (one to seven) to come up to full light output. Also, if power to the lamp is lost or turned off, the arc tube must cool before the arc can be restruck and light produced. Up to seven minutes (for mercury vapor lights) may be required.

MERCURY VAPOR LAMPS

The mercury vapor (MV) lamp produces light when electrical current passes through a small amount of mercury vapor. The lamp consists of two glass envelopes: an inner envelope in which the arc is struck, and an outer or protective envelope. The mercury vapor lamp, like the fluorescent lamp, requires a ballast designed for its specific use.

Although, used extensively in the past, mercury vapor lamps are not as popular as other HID sources today due to its relatively low efficacy. However, because of their low cost and long life (16,000 to 24,000 hours), mercury vapor lamps still find some applications.

The color rendering qualities of the mercury vapor lamp are not as good as those of incandescent lamps. A significant portion of the energy radiated is in the ultraviolet region resulting in a "bluish" light in the standard lamp. Through use of phosphor coatings on the inside of the outer envelope, some of the energy is converted to visible light resulting in better color rendition and use in indoor applications.

Mercury vapor lamp sizes range from 40 to 1,000 watts.

METAL HALIDE LAMPS

The metal halide (MH) lamp is very similar in construction to the mercury vapor lamp. The major difference is that the metal halide lamp contains various metal halide additives in addition to the mercury vapor. The efficacy of metal halide lamps is from 1.5 to 2 times that of mercury vapor lamps. The metal halide lamp produces a relatively "white" light, equal or superior to presently available mercury vapor lamps. The main disadvantage of the metal halide lamp is its relatively short life (7,500 to 20,000 hours).

Metal halide lamp sizes range from 175 to 1,500 watts. Ballasts designed specifically for metal halide lamps must be used.

HIGH PRESSURE SODIUM LAMPS

The high pressure sodium (HPS) lamp has the highest efficacy of all lamps normally used indoors. It produces light when electricity passes through a sodium vapor. This lamp also has two envelopes, the inner one being made of a polycrystalline aluminum in which the light-producing arc is struck. The outer envelope is protective, and may be either clear or coated. The light produced by this lamp is a "golden-white" color.

Although the HPS lamp first found its principal use in outdoor lighting, it now is a readily accepted light source indoors in industrial plants. It also is being used in many commercial and institutional applications as well.

HPS lamp size ranges from 35 to 1,000 watts. Ballasts designed specifically for high pressure sodium lamps must be used.

LOW PRESSURE SODIUM LAMPS

The low pressure sodium (LPS) is the most efficient light source presently available, providing up to 183 lumens per watt. The light in this lamp is produced by a U-shaped arc tube containing a sodium vapor. Its use indoors is severly restricted, however, because it has a monochromatic (yellow) light output. Consequently, most colors illuminated by this light source appear as tones of gray.

Low pressure sodium lamps range in size from 18 watts to 180 watts. Ballasts designed specifically for LPS must be used. The primary use of these lamps is street lighting as well as outdoor area and security lighting. Indoor applications such as warehouses are practical where color is not important.

LUMINAIRE EFFICIENCY

In the previous section, it was seen that a lamp produces an amount of light (measured in lumens) which depends on the power consumed and the type of lamp. Equally important to the amount of light produced by a lamp, is the amount of light which is "usable" or provides illumination for the desired task. Luminaires, or lighting fixtures, are used to direct the light to a usable location, dependent on the specific requirements of the area to be lighted. Regardless of the luminaire type, some of the light is directed in non-usable directions, is absorbed by the luminaire itself or is absorbed by the walls, ceiling or floor of the room.

The coefficient of utilization, or CU, is a factor used to determine the efficiency of a fixture in delivering light for a specific application. The coefficient of utilization is determined as a ratio of light output from the luminaire that reaches the workplane to the light output of the lamps alone. Luminaire manufacturers provide CU data in their catalogs which are dependent on room size and shape, fixture mounting height and surface reflectances. Table 6-2 illustrates the form in which a vendor summarized the data used for determining the coefficient of utilization.

To determine the coefficient of utilization, the room cavity ratio, wall reflectance, and effective ceiling cavity reflectance must be known.

Most data assumes a 20% effective floor cavity reflectance. To determine the coefficient of utilization, the following steps are needed:

(a) Estimate wall and ceiling reflectances.

(b) Determine room cavity ratio.

(c) Determine effective ceiling reflectance (pCC).

Step (a)
Typical reflectance values are shown in Table 6-3.

Table 6-2. Vendor Data for 175 Watt Mercury Vapor Lamp—Medium Spread Deflector

Coefficients of Utilization/Effective Floor Cavity Reflectance 20% (pFC)

% REFLECTANCE EFF. CEIL. (pCC)	WALL (pW)	ROOM CAVITY RATIO									
		1	2	3	4	5	6	7	8	9	10
80	50	0.854	0.779	0.711	0.647	0.591	0.539	0.490	0.446	0.407	0.355
	30	0.828	0.739	0.664	0.594	0.533	0.481	0.432	0.388	0.349	0.296
	10	0.805	0.705	0.626	0.552	0.491	0.440	0.392	0.347	0.309	0.258
70	50	0.832	0.761	0.698	0.635	0.578	0.530	0.483	0.438	0.401	0.349
	30	0.808	0.724	0.653	0.585	0.526	0.475	0.426	0.384	0.345	0.295
	10	0.786	0.695	0.618	0.546	0.486	0.434	0.387	0.344	0.308	0.256
50	50	0.788	0.725	0.669	0.610	0.558	0.511	0.466	0.424	0.388	0.339
	30	0.770	0.696	0.632	0.568	0.513	0.464	0.416	0.375	0.338	0.288
	10	0.754	0.670	0.602	0.534	0.478	0.428	0.382	0.339	0.303	0.253
30	50	0.750	0.694	0.642	0.587	0.539	0.495	0.450	0.412	0.377	0.329
	30	0.736	0.671	0.612	0.552	0.499	0.453	0.408	0.367	0.331	0.282
	10	0.722	0.649	0.586	0.523	0.469	0.421	0.375	0.335	0.299	0.249
10	50	0.716	0.665	0.618	0.566	0.521	0.479	0.438	0.399	0.366	0.319
	30	0.704	0.645	0.592	0.536	0.487	0.442	0.399	0.360	0.325	0.276
	10	0.693	0.628	0.571	0.511	0.460	0.413	0.370	0.330	0.294	0.245

Table 6-3. Typical Reflection Factors

COLOR	REFLECTION FACTOR
White and very light tints	.75
Medium blue-green, yellow or gray	.50
Dark gray, medium blue	.30
Dark blue, brown, dark green, and wood finishes	.10

Steps (b) and (c)

Once the wall and ceiling reflectances are estimated it is necessary to analyze the room configuration to determine the effective reflectances. Any room is made up of a series of cavities which have effective reflectances with respect to each other and the work plane. Figure 6-1 indicates the basic cavities.

Figure 6-1. Cavity Configurations

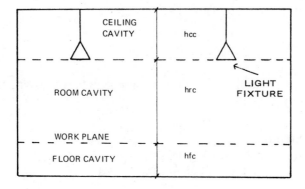

The space between fixture and ceiling is the ceiling cavity. The space between the work plane and the floor is the floor cavity. The space between the fixture and the work plane is the room cavity. To determine the cavity ratio use Figure 6-1 to define the cavity depth and then use Formula 6-1.

(Formula 6-1) $\text{Cavity Ratio} = \dfrac{5 \times d \times (L+W)}{L \times W}$

Where d = depth of the cavity as defined in Figure 6-1
L = Room (or area) length
W = Room (or area) width

To determine the effective ceiling or floor cavity reflectance, proceed in the same manner to define the ceiling or floor cavity ratio, then refer to Table 6-4 to find the corresponding effective ceiling or floor cavity reflectance.

SIM 6-1

For Process Plant No. 1, determine the coefficient of utilization for a room which measures $24' \times 100'$. The ceiling is $20'$ high and the fixture is mounted $4'$ from the ceiling. The tasks in the room are performed on work benches $3'$ above the floor. Use the data in Table 6-2.

Answer

Step (a)
Since no wall or ceiling reflectance data was given, assume a ceiling of .70 and wall of .5.

Step (b)
Assume $3'$ working height.
hrc = 20-4-3 = 13 (from Figure 6-1)
From Formula 6-1, RCR = 3.4

Step (c)
From Figure 6-1, hcc = 4
From Formula 6-1, CCR = 1
From Table 6-4, pCC = 58
From Table 6-2, Coefficient of Utilization = 0.64
     (interpolated)

Table 6-4. Per Cent Effective Ceiling or Floor Cavity Reflectance for Various Reflectance Combinations

% Ceiling or Floor Reflectance	90				80				70			50			30				10		
% Wall Reflectance → / Ceiling or Floor Cavity Ratio ↓	90	70	50	30	80	70	50	30	70	50	30	70	50	30	65	50	30	10	50	30	10
0	90	90	90	90	80	80	80	80	70	70	70	50	50	50	30	30	30	30	10	10	10
0.1	90	89	88	87	79	79	78	78	70	69	68	50	49	48	30	30	30	30	10	10	10
0.2	89	88	86	85	79	78	77	76	68	67	66	49	48	47	30	29	29	28	10	10	9
0.3	89	87	85	83	78	77	75	74	68	66	64	49	47	46	30	29	29	27	10	10	9
0.4	88	86	83	81	78	76	74	72	67	65	63	48	46	45	30	29	28	26	11	10	9
0.5	88	85	81	78	77	75	73	70	66	64	61	48	46	44	29	28	27	25	11	10	9
0.6	88	84	80	76	77	75	71	68	65	62	59	47	45	43	29	28	26	25	11	10	9
0.7	88	83	78	74	76	74	70	66	65	61	58	47	44	42	29	28	26	24	11	10	8
0.8	87	82	77	73	75	73	69	65	64	60	56	46	43	41	29	27	25	23	11	10	8
0.9	87	81	76	71	75	72	68	63	63	59	55	46	43	40	29	27	25	22	11	9	8
1.0	86	80	74	69	74	71	66	61	63	58	53	46	42	39	29	27	24	22	11	9	8
1.1	86	79	73	67	74	70	65	60	62	57	52	45	41	38	29	26	24	21	11	9	8
1.2	86	78	72	65	73	69	64	58	61	56	50	45	41	37	29	26	23	20	12	9	7
1.3	85	78	70	64	73	68	63	57	61	55	49	45	40	36	28	26	23	20	12	9	7
1.4	85	77	69	62	72	68	62	55	60	54	48	44	40	35	28	25	22	19	12	9	7
1.5	85	76	68	61	72	67	61	54	59	53	47	44	39	34	28	25	22	18	12	9	7
1.6	85	75	66	59	71	67	60	53	59	52	45	44	39	33	28	25	21	18	12	9	7
1.7	84	74	65	58	71	66	59	52	58	51	44	43	38	32	28	25	21	17	12	9	7
1.8	84	73	64	56	70	65	58	50	57	50	43	43	37	32	28	25	21	17	12	9	6
1.9	84	73	63	55	70	65	57	49	57	49	42	43	37	31	28	25	20	16	12	9	6
2.0	83	72	62	53	69	64	56	48	56	48	41	43	37	30	28	24	20	16	12	9	6

(more)

6	6	6	6	6	5	5	5	5	5	5	5	5	5	5	5	4	4	4	4	4	4	4	4	4	4	4	4	4	4
9	9	9	9	9	9	9	9	9	8	8	8	8	8	8	8	8	8	8	8	8	8	8	8	8	8	8	8	8	8
13	13	13	13	13	13	13	13	13	13	13	13	13	13	13	13	13	13	13	13	13	13	13	13	14	14	14	14	14	14
16	15	15	14	14	13	13	13	12	12	12	11	11	11	11	10	10	10	10	9	9	9	9	8	8	8	8	8	7	7
20	19	19	19	18	18	18	18	17	17	17	16	16	16	16	15	15	15	15	15	14	14	14	14	14	14	13	13	13	13
24	24	24	24	23	23	23	23	23	22	22	22	22	22	22	21	21	21	21	21	21	20	20	20	20	20	20	19	19	19
28	28	28	28	27	27	27	27	27	27	27	27	27	27	26	26	26	26	26	26	26	26	26	26	25	25	25	25	25	25
29	29	28	27	27	26	26	25	25	24	24	23	23	22	22	21	21	21	20	20	20	19	19	19	19	18	18	18	18	17
36	36	35	35	34	34	33	33	33	32	32	31	31	31	30	30	30	29	29	29	28	28	28	27	27	27	26	26	26	26
43	42	42	42	41	41	41	41	40	40	40	40	39	39	39	39	38	38	38	38	37	37	37	37	37	36	36	36	36	36
40	39	38	37	36	35	34	33	33	32	31	30	30	29	29	28	27	27	26	26	25	25	25	24	24	24	23	23	23	22
47	46	46	45	44	43	43	42	41	40	40	39	39	38	38	37	37	36	36	35	35	34	34	34	33	33	33	32	32	32
56	55	54	54	53	53	52	52	51	51	50	50	49	49	48	48	48	47	47	46	46	46	45	45	45	44	44	44	44	43
47	45	44	43	42	41	40	39	38	38	37	36	35	34	33	33	32	31	30	30	29	29	28	28	27	26	26	25	25	25
55	54	53	52	51	50	49	48	48	47	46	45	44	44	43	42	42	41	40	40	39	39	38	38	37	37	36	36	35	35
63	63	62	61	61	60	60	59	58	58	57	57	56	56	55	54	54	53	53	52	52	51	51	51	50	50	49	49	49	48
69	69	68	67	67	66	66	66	65	65	64	64	64	63	63	62	62	62	61	61	60	60	60	59	59	59	58	58	58	57
52	51	50	48	47	46	45	44	43	42	41	40	39	38	37	36	35	35	34	33	32	32	31	30	30	29	29	28	28	27
61	60	59	58	57	56	55	54	53	52	51	50	49	48	48	47	46	45	45	44	43	43	42	41	41	40	40	39	38	38
71	70	69	68	68	67	66	66	65	64	64	63	62	62	61	60	60	59	59	58	57	57	56	56	55	55	54	54	53	53
83	83	83	82	82	82	82	81	81	81	80	80	80	80	79	79	79	79	78	78	78	78	78	77	77	77	77	76	76	76
2.1	2.2	2.3	2.4	2.5	2.6	2.7	2.8	2.9	3.0	3.1	3.2	3.3	3.4	3.5	3.6	3.7	3.8	3.9	4.0	4.1	4.2	4.3	4.4	4.5	4.6	4.7	4.8	4.9	5.0

Ceiling or Floor Cavity Ratio

LIGHT LOSS FACTOR

The amount of light produced by a luminaire as determined by the lamp lumen output and the fixture coefficient of utilization is the initial value only. Over time the light reaching the task surface will depreciate due to two factors collectively known as the light loss factor (LLF).

The light loss factor (LLF) takes into account that the lumen output of all lamps depreciates with time (LLD) and that the lumen output depreciates due to dirt build-up on the lamp and fixture (LDD). Formula 6-2 illustrates the relationship of these factors.

(Formula 6-2)   LLF = LLF x LLD

To reduce the number of lamps required which in turn reduces energy consumption, it is necessary to increase the overall light loss factor. This is accomplished in several ways. One is to choose a luminaire which minimizes dust build-up. The second is to improve the maintenacne program to replace lamps prior to burn-out. Thus if it is known that a group relamping program will be used at a given percentage of rated life, the appropriate lumen depreciation factor can be found. It may be decided to use a shorter relamping period in order to increase (LLD) even further.

Figure 6-2 illustrates the effect of lumen depreciation and dirt build-up for a typical luminaire. Manufacturer's data should be consulted when estimating LLD and LDD for a luminaire.

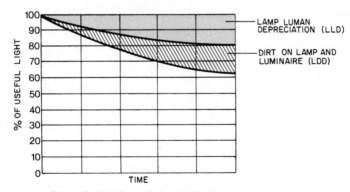

Figure 6-2. Light Output Reduction With Time

ILLUMINATION LEVELS

The amount of light that illuminates a surface is measured in lumens per square foot or footcandles. Table 6-3 shows selected illumination level ranges as recommended by the Illuminating Engineering society in the 1981 Lighting Handbook. Note that these values are recommended for the performance of a specific task and that a room with various task areas would have various recommended illumination levels.

The values in Table 6-3 are intended as guidelines only. The age of the occupants, the inherent difficulty in viewing the object, the importance of speed and/or accuracy for visual performance and the reflectance of the task must be considered when applying these illumination ranges (see IES Handbook for further information).

Table 6-3. Recommended Illumination Values for Selected Areas

Area	Activity	Illuminance Range (Footcandles) on Task
Industrial Assembly	Simple	20-50
	Difficult	100-200
	Extracting	500-1000
Drafting	High Contrast Media	50-100
	Low Contrast Media	100-200
Food Service Facilities	Dining	5-10
Machine Shops	Rough Bench Work	20-50
	Fine Bench Work	200-500
Offices	Lobby	10-20
	Conference Room	20-50
Parking	Open Area	0.5-2
	Closed Area	5-10
Reading	Ditto Copy	50-100
	Ball Point Pen	20-50
	#3 Pencil and Softer Leads	50-100
	Typed Originals	20-50
	Glossy Magazines	20-50

THE LUMEN METHOD

Combining the concepts presented in the previous sections, we can use Formula 6-3 to determine the number of lamps to provide average, uniform lighting levels. This formula is known as the lumen method.

(Formula 6-3) $$N = \frac{E \times A}{Lu \times LLF \times CU}$$

where

N is the number of lamps required
E is the required illuminance in footcandles
A is the area of the room in square feet
Lu is the lumen output of the lamp
LLF is the light loss factor which accounts for lamp lumen depreciation and lamp (and fixture) dirt depreciation
CU is the coefficient of utilization

Note that for a specific area and level of illumination for the area, the only means that the lighting designer has for reducing the number of lamps (and consequently the power consumption) required is to use the highest values of Lu, CU and LLF.

SIM 2

For the situation described in Example Problem 1, determine the required number of fixtures to give an average, maintained footcandle level of 50. The light loss factor is estimated to be 0.7. The lamps are 175 watt mercury vapor with one lamp per fixture and an initial lumen output of 8,500.

Answer

No. of fixtures $= \dfrac{\text{Area} \times \text{Desired Maintained Footcandle}}{\text{Lumens} \times CU \times LU}$

$= \dfrac{24 \times 100 \times 50}{8,500 \times 0.64 \times 0.7}$ = 31.5 or 32 fixtures

THE POINT METHOD

The lumen method is useful in determining the average illumination in an area but sometimes it is desirable to know the illumination level due to one or more lighting fixtures upon a specified point within the area.

The point method (Formula 6-4) computes the level of illumination in footcandles by determining the contribution of a single light source in the area. For multiple light sources, Formula 6-4 must be used for each one and the results summed. Reflections from walls, ceilings and floors are not considered in this method, consequently it is especially useful for very large areas, outdoor lighting and areas where room surfaces are dark or dirty. Additionally, the formula holds true for point sources only. Caution must be exercised when using the point method for fluorescent sources or for luminaires with large reflectors. As a rule of thumb, if the maximum dimensions of the source are no more than one-fifth the distance to the point of calculation, the source will be considered a point source and the calculated illumination will be reasonably accurate.

(Formula 6-4)   Horizontal Footcandles = $\dfrac{cp \times h}{d^3}$
                            on a Task

where

cp is the candlepower at the desired angle Θ (see Figure 6-3) obtained from manufacturer's data (see Figure 6-4)

h is the height of the fixture above the horizontal plane of the task

d is the distance from the light source to the task

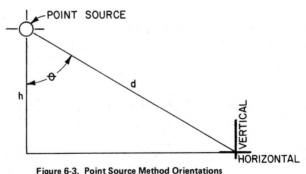

Figure 6-3. Point Source Method Orientations

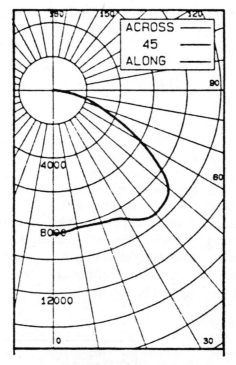

Figure 6-4. Candlepower Distribution for
A 400 Watt HPS Low Bay Fixture

A candlepower distribution curve (see figure 6-4) shows the luminous intensity of a fixture (measured in caldelas) for a range of angular orientations to the fixture. (0° is taken as directly beneath the fixture.)

SIM 3

What is the illumination on a surface due to a single 400 watt high pressure sodium light source represented by the data in Figure 6-4 which is 10' in horizontal distance from the workplane? The vertical distance above the workplane (h) is 10'.

Answer

From trigonometric relationships, the angle Θ = arctangent 10/10 = 45°. From Figure 6-4 this gives a candlepower of

9500 candelas. d is found from trigonometric relationships to be h/cos Θ = 10/cos 45° = 14.1'. Therefore,

$$\text{Footcandles} = \frac{9500 \times 10}{(14.1)^3} = 33.9$$

FIXTURE LAYOUT

The fixture layout is dependent on the area. The initial layout should have equal spacing between lamps, rows and columns. The end fixture should be located at one-half the distance between fixtures. The maximum distance between fixtures usually should not exceed the mounting height unless the manufacturer specifies otherwise. Figure 6-5 illustrates a typical layout. If the fixture is fluorescent, it may be more practical to run the fixtures together. Since the fixtures are 4 feet or 8 feet long, a continuous wireway will be formed.

**Figure 6-5
Typical Fixture
Layout**

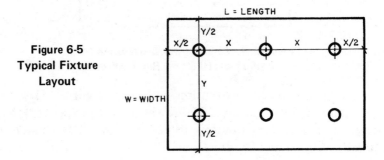

SIM 6-4

For SIM 6-2 design a lighting layout.

Answer

From SIM 6-2, thirty-two 175 watt mercury vapor lamps are required.

		Rows	Columns	X Spacing	Y Spacing
Typical					
Combinations	(a)	4	8	12.5	6
	(b)	3	11	9	8
	(c)	2	16	6	12

(a) $8X = 100$ (b) $11X = 100$ (c) $16X = 100$

     $X = 12.5$ $X = 9$ $X = 6.2$

     $4Y = 24$ $3Y = 24$ $2Y = 24$

     $Y = 6$ $Y = 8$ $Y = 12$

Alternate (b) is recommended even though it requires one more fixture. It results in a good layout, illustrated following.

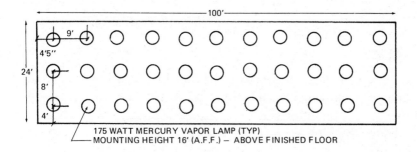

175 WATT MERCURY VAPOR LAMP (TYP)
MOUNTING HEIGHT 16' (A.F.F.) — ABOVE FINISHED FLOOR

CIRCUITING

Number of Lamps Per Circuit

A commonly used circuit loading is 1600 watts per lighting circuit breaker. This load includes fixture voltage and ballast loss. In SIM 6-4, assuming a ballast loss of 25 watts per fixture, a 20 amp circuit breaker, and #12 guage wire, eight lamps could be fed from each circuit breaker. A single-phase circuit panel is illustrated in Figure 6-6. (Note: In practice ballast loss should be based on manufacturer's specifications.)

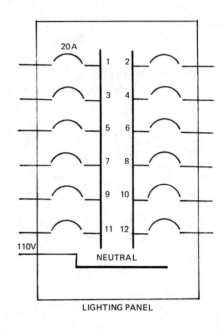

**Figure 6-6
Single-Phase
Circuit Panel**

LIGHTING PANEL

SIM 6-5

Next to each lamp place the panel designation and circuit number from which each lamp is fed; i.e., A-1, A-2, etc.

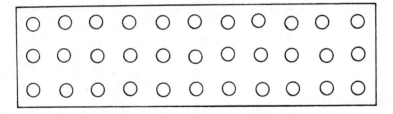

Answer

A-1	A-1	A-1	A-1	A-1	A-1	A-1	A-1	A-4	A-4	A-4
○	○	○	○	○	○	○	○	○	○	○
A-2	A-2	A-2	A-2	A-2	A-2	A-2	A-2	A-4	A-4	A-4
○	○	○	○	○	○	○	○	○	○	○
A-3	A-3	A-3	A-3	A-3	A-3	A-3	A-3	A-4	A-4	A-5
○	○	○	○	○	○	○	○	○	○	○

SIM 6-6

Designate a hot line from the circuit breaker with a small stroke and use a long stroke as a neutral; i.e., —⫻⫻— 4 wires, 2 hot and 2 neutrals. The lamps are connected with conduit as shown below. Designate the hot and neutrals in each branch.

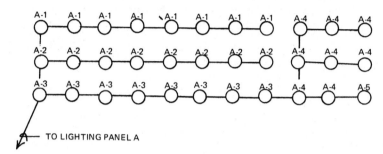

Hint—start wiring from the last fixture in the circuit.

Answer

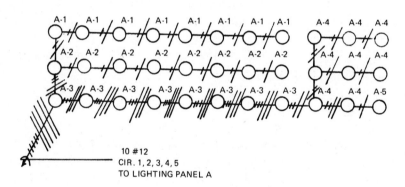

Note that only a single neutral wire is required for each 3 different phase wires.

POINTS ON LIGHTING DRAWINGS

• Choose a lighting drawing scale based on the area to be lighted and the detail required. Typical drawing scale: 1/8″ equals one foot.

- Identify all symbols for lighting fixtures.
- Include circuit numbers on all lights.
- Include a note on fixture mounting height.
- Show "homerun" to lighting panels. "Homerun" indicates the number of wires and conduit size from the last outlet box.
- Use notes to simplify drawing. For example: All wires shall be 2 #12 in 3/4" conduit unless otherwise indicated. Remember the information put on a drawing or specification should be clear to insure proper illumination.

LIGHTING QUALITY

Illumination levels calculated by the lumen and point methods at best give only a "ballpark" estimate of the actual footcandle value to be realized in an installation. Many inaccuracies can be present including: differences between rated lamp lumen output and actual values; difficulty in predicting actual light loss factors; difficulty in predicting room surface reflectances; inaccurate CU information from a manufacturer; non-rectangular shaped rooms.

Precise illumination levels are not critically important, however. Of equal importance to lighting quantity is lighting quality. Very few people can perceive a difference of plus or minus ten footcandles, but poor quality lighting is readily apparent to anyone and greatly affects our ability to comfortably "see" a task.

Of the many factors affecting the quality of a lighting installation, glare has the greatest impact on our ability to comfortably perceive a task. Figure 6-7 shows the two types of glare normally encountered.

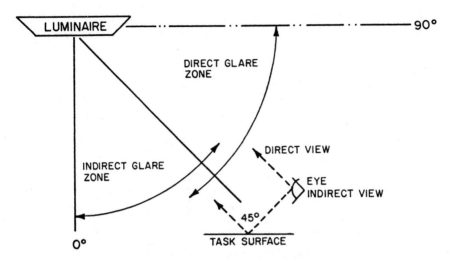

Figure 6-7. Direct and Indirect Glare Zones

DIRECT GLARE

Direct glare is often caused by a light source in the midst of a dark surface. Direct glare also can be caused by light sources, including sunlight, in the worker's line of sight. A rating system has been developed for assessing direct glare called Visual Comfort Probability (VCP). The VCP takes into account fixture brightness at different angles of view, fixture size, room size, fixture mounting height, illumination level, and room surface reflectances.

Most manufacturers publish VCP tables for their fixtures. A VCP value of 70 or higher usually provides acceptable brightness for an office situation. Table 6-5 shows a typical VCP table.

Room Size (in ft.)		Luminaires Lengthwise		Luminaires Crosswise	
		Ceiling Height (in feet)			
W	L	8.5	10.0	8.5	10.0
20 x	20	75	72	73	70
	30	75	72	73	70
	40	75	73	73	71
	60	75	73	72	71
30 x	20	78	75	77	73
	30	78	75	76	73
	40	77	75	75	72
	60	76	74	74	72
	80	76	74	73	72
40 x	20	81	78	80	77
	30	79	77	78	76
	40	78	77	76	75
	60	77	76	75	74
	80	77	75	74	73
	100	76	75	74	73
60 x	30	81	79	80	77
	40	79	78	78	76
	60	78	77	76	74
	80	77	76	75	74
	100	77	76	75	73
100 x	40	81	80	80	78
	60	80	78	78	76
	80	78	77	77	75
	100	78	76	76	74

**Wall Reflectance, 50%
Ceiling Cavity Reflectance, 80%
Floor Cavity Reflectance, 20%
Work Plane Illumination, 100fc

Table 6-5.

INDIRECT GLARE

Indirect glare occurs when light is reflected off of a surface in the work area. When the light bounces off a task surface, details of the task surface become less distinct because contrast between the foreground and background, such as the type on this page and the paper on which it is printed, is reduced. This is most easily visualized if a mirror is placed at the task surface and the image of a light fixture is seen at the normal viewing angle.

This form of indirect glare is called a *veiling reflection* because its effects are similar to those that would result were a thin veil placed between the worker's eyes and the task surface. Veiling reflections can be reduced by:

1) Orient fixtures (or work surface) so that the light produced is not in the indirect glare zone (generally to the side and slightly behind the work position gives the best results).

2) Select fixtures which direct the light above the worst veiling angles (generally 30° or greater). These fixtures have "batwing" distribution patterns such as shown in Figure 6-4. Note, that in selecting fixtures to minimize indirect glare, care must be taken not to select fixtures that are a source of excessive direct glare.

ENERGY CONSERVATION CONSIDERATIONS

Recent concern for energy conservation has focused attention on lighting as an area for potential savings since it can account for 25% of total energy use in an office building. This attention has resulted in many new products which greatly decrease the amount of energy needed for lighting. Unfortunately, blanket application of energy conserving techniques has also resulted in some poorly lit applications which save energy at the expense of worker productivity.

LAMP/LUMINAIRE EFFICIENCIES

As noted in the previous sections, there are wide variations in the efficacies of light sources. By selecting the most efficient light source within the color and room configuration constraints, significant energy savings are possible. Additionally, a new generation of "energy efficient" fluorescent lamps and ballasts are available which offer 5-20% savings over their standard counterparts.

Also, as shown in the previous sections, the choice of luminaire can have a great impact on the energy used for lighting since it determines how much light reaches the task.

NON-UNIFORM LIGHTING

The lumen method presented previously is useful for calculating average uniform illumination for an area, but illumination levels presented in Table 6-3 are for specific tasks in an area. By tailoring illumination levels to the various tasks in an area, significant energy savings are possible.

For example, if 30% of an office area is comprised of desks at which people will be reading material written in pencil, the recommended illumination level is from 50 to 100 footcandles depending on the application. In the remaining 70% of the office, however, if the area is used for general passageways or as a lobby area, the required illumination level is only 10 to 20 footcandles. By directing high levels of illumination only to the task areas, significant energy savings are possible. (Note, that it is generally recommended to limit the ratio of task levels to non-task levels to 3 to 1 to minimize fatigue caused by excessive contrast.)

This technique is referred to as non-uniform or task-ambient lighting. It can be accomplished by positioning fixtures over the task location or by providing a small fixture which is mounted on the desk or machine tool, for example, to provide localized lighting.

GROUP RELAMPING

As seen in Figure 6-2 light output of a fixture depreciates over time due to lamp aging and dirt accumulation. If a minimum footcandle level is desired, initial footcandles must be as much as 40% higher than that desired. Consequently, many more fixtures and consequent power consumption is required to compensate for these depreciation factors.

If a systematic program is initiated, however, to periodically clean the fixtures and relamp before the end of rated life, the number of fixtures can be reduced while maintaining the desired illumination level. This technique is known as group relamping and lighting maintenance. It has the effect of raising the light loss factor (LLF) in Formula 6-3.

LEVEL CONTROLS

Areas with daylight available through windows and skylights can achieve significant energy savings by reducing the lighting system output to maintain a desired illumination level. This can be accomplished by either turning the fixture off, by reducing its output by switching off some of the lamps in the fixture or by using special dimming circuitry.

Additionally, level controls can reduce lighting system energy consumption during "non-production" times when lights are needed. For example, an office which required 70 footcandles during business hours only required 20 footcandles for cleaning at night. By reducing the light levels to 20 footcandles during the cleaning periods, significant energy savings are possible.

Also, level controls (specifically dimming controls) can be used to compensate for light depreciation factors thereby providing required footcandle levels with the minimum possible power consumption.

ON/OFF CONTROLS

One of the simplest and most effective means of controlling lighting energy consumption is by turning off the lights when not needed. To effectively accomplish this may require the addition of switches in each office, grouping of lights into "zones" of usage types, the use of occupancy sensors (either ultrasonic or infrared) to detect when occupants are present and/or the use of an energy management system to automatically schedule lighting operation. Energy management systems are explained in Chapter 11.

JOB SIMULATION–SUMMARY PROBLEM

Job 4

(a)   The Ajax Plant, Job 1, Chapter 2, contains a workshop area with an area of 20' X 18'6".

For this area compute the number of lamps required, the space between fixtures, and the circuit layout. Use two 40-watt fluorescent lamps per fixture, 2900 lumens per lamp, light loss factor = .7, 110-volt lighting system, 20-watt ballast loss per fixture, and a fixture length of 2' X 4'. Use luminaire data of Table 6-6, ceiling height 20' and a desired footcandle level of 40.

Table 6-6. Coefficient of Utilization
20% Effective Floor Cavity Reflectance

Effective Ceiling Cavity Reflectance	80%			50%		
Wall Reflectance	50	30	10	50	30	10
RCR						
10	.33	.26	.22	.31	.26	.22
9	.43	.35	.27	.40	.35	.29
8	.58	.42	.35	.48	.42	.36
7	.58	.50	.42	.55	.48	.42
6	.64	.57	.49	.61	.54	.47
5	.72	.65	.59	.65	.60	.56
4	.77	.71	.64	.71	.65	.60
3	.82	.76	.70	.74	.69	.63
2	.87	.82	.77	.78	.74	.70
1	.91	.87	.83	.81	.78	.75
Spacing not to exceed 1 X Mounting Height						

Analysis

The area of the workshop is $18'6'' \times 20'$.

Assume   hfc = 3

            hcc = 3

Therefore,   hrc = 14

Assume 70% ceiling reflectance

       50% wall reflectance

The room cavity ratio is 7 and the effective ceiling cavity ratio pcc = 53. Thus C.U. = .55.

$$\text{No. of Fixtures} = \frac{20' \times 18\frac{1}{2}' \times 40}{2 \times 2900 \times .55 \times .70} = 7$$

Each 20 amp lighting circuit can provide power for up to 16 fixtures.

Layout Spacing

$3x = 20$
$x = 6.6$
$3y = 18\frac{1}{2}'$
$y = 6'2''$

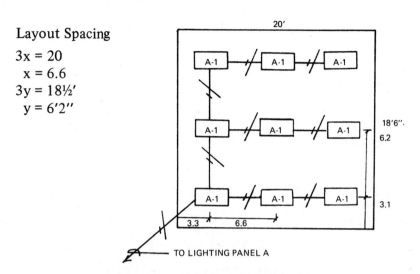

Note: With emphasis on energy conservation, a lighting layout using 6 fixtures may be preferable.

SUMMARY

Energy conservation is influencing lighting design. Increased emphasis is being placed on minimizing lighting energy use by using lamps and luminaries which have high lumen outputs and coefficients of utilization. Today's lighting systems incorporate switching and automatic control devices to make it easy to turn off lights when they are not required. Lighting systems need to be analyzed on a first and operating cost basis to insure that the increasing energy costs are taken into account.

Lighting system design must not only consider the quantity of illumination but also the quality of illumination. The choice of a luminaire and its location play an important part in comfortably perceiving a task. An awareness of the importance of quality lighting can result in a visual environment which is productive as well as energy efficient.

Chapter 7

USING LOGIC TO
SIMPLIFY CONTROL SYSTEMS

Automated process plants are controlled by electrical hardware specified on elementary diagrams and designed by electrical engineers. Standard inexpensive contactors, timing devices, relays and other simple electro-mechanical devices provide control for practically every circuit. In this chapter an approach for analyzing and designing elementary diagrams is developed. Learn the logic language which will form a new way for communicating.

ELECTRICAL HARDWARE

Electrical hardware used for control includes:

• *Relays* — Devices which when energized close. Contacts physically located on the relay will either open or close when the relay is energized.

• *Timers and time delay relays* — Devices whose contacts close or open after a preset time.

• *Pushbuttons* — Devices which are used to actuate a control system; i.e., Stop-Start pushbuttons.

• *Programmers* — Devices whose contacts open and close in a preset sequence.

Present trend is toward plug-in type relays, prepackaged solid-state components.

SYMBOLS

Symbols commonly used in electrical control schematics are illustrated in Figure 7-1. These symbols are based on the Joint Industrial Council (JIC) Standards.

THE ELECTRICAL ELEMENTARY (SCHEMATIC)

Figure 7-2 illustrates a typical electrical elementary diagram. Notice that line identifications on the left are used as references to locate relay contacts on the right. For example, relay R_1 has a normally open contact on line six.

Steps for Analyzing Electrical Control Circuits

• Look at one line of operation at a time.

• Trace a path of power from left to right. Every contact to the left of the electrically operated line must be closed for the device to operate.

• Each line is identified with a consecutive number on the left.

• Numbers at the right of the line next to a relay show on what lines the device has contacts. Underlined numbers indicate a normally closed contact.

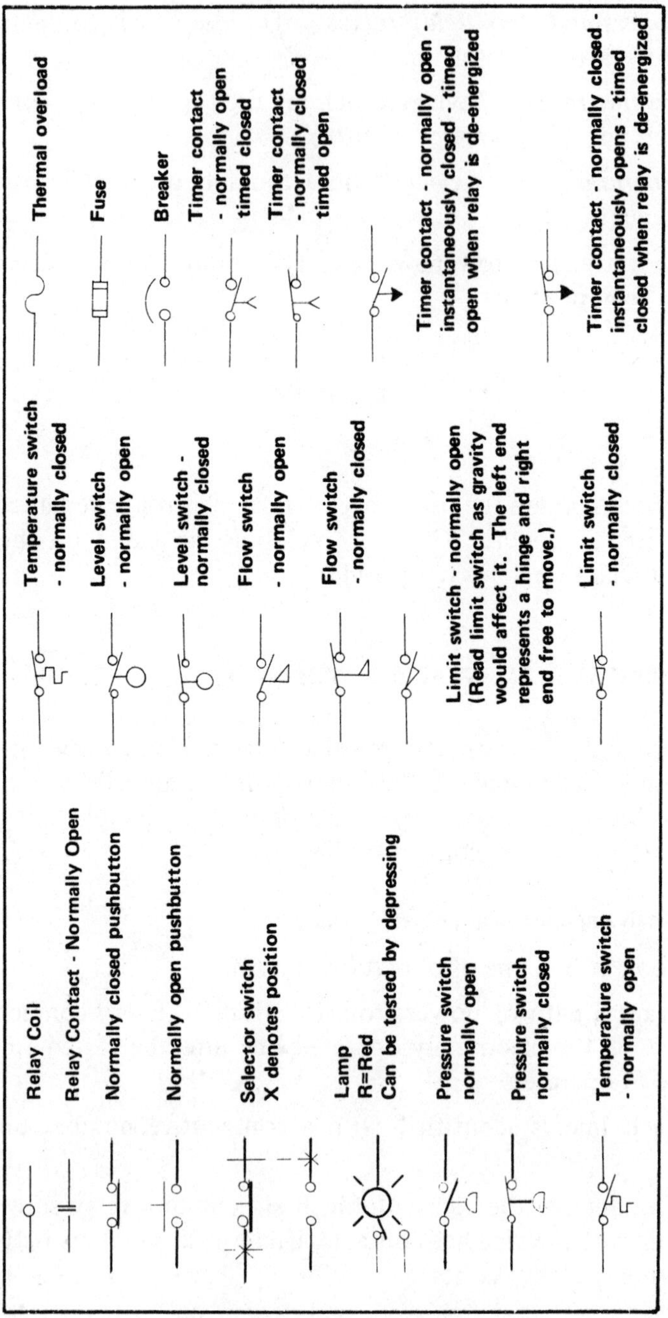

Figure 7-1. Electrical Symbols

Relay Coil

Relay Contact - Normally Open

Normally closed pushbutton

Normally open pushbutton

Selector switch
X denotes position

Lamp
R=Red
Can be tested by depressing

Pressure switch
normally open

Pressure switch
normally closed

Temperature switch
- normally open

Temperature switch
- normally closed

Level switch
- normally open

Level switch -
normally closed

Flow switch
- normally open

Flow switch
- normally closed

Limit switch - normally open
(Read limit switch as gravity
would affect it. The left end
represents a hinge and right
end free to move.)

Limit switch
- normally closed

Thermal overload

Fuse

Breaker

Timer contact
- normally open
timed closed

Timer contact
- normally closed
timed open

Timer contact - normally open -
instantaneously closed - timed
open when relay is de-energized

Timer contact - normally closed -
instantaneously opens - timed
closed when relay is de-energized

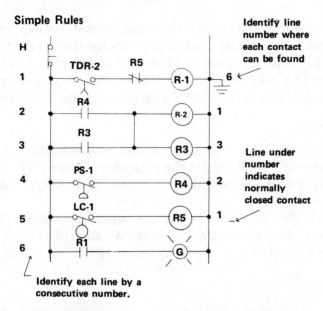

Figure 7-2. Format of an Electrical Elementary Diagram

Simple Control Schemes

In order to control a motor, a starter is required. A typical motor elementary diagram, including the power portion, is illustrated in Figure 7-3. HCA represents the holding coil or contactor of the starter. HCA is simply a relay which has contacts capable of interrupting power to the associated motor.

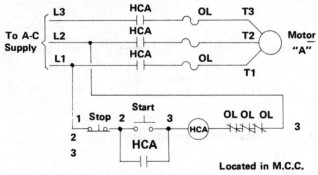

Stop Start - Full Voltage
Non-reversing Motor

Figure 7-3. Elementary Diagram for a FVNR Motor

If the stop-start pushbutton were located locally, 3 wires, numbers 1, 2, 3 would need to run from the field to the M.C.C.

The control voltage in Figure 7-3 is 480 volts, but if the circuit had a solenoid valve, limit switch, etc., 110 volts would be required. (Control devices are usually rated for 110 volts.)

Interlocking

Two typical interlocking schemes are illustrated in Figure 7-4. Scheme "A" illustrates the case where Motor "B" can not be started unless Motor "A" is running. If Motor "A" stops so will Motor "B".

Scheme "B" illustrates the case where once Motor "B" is running it does not matter if Motor "A" stops. The permissive interlock is only required to start Motor "B".

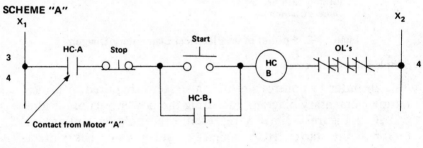

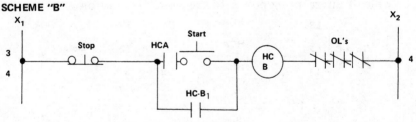

Figure 7-4. Two Interlocking Schemes

Reversing Motors

To change the rotation of a motor it is necessary to switch any two of the motor leads. Figure 7-5 illustrates the elementary for a reversing motor.

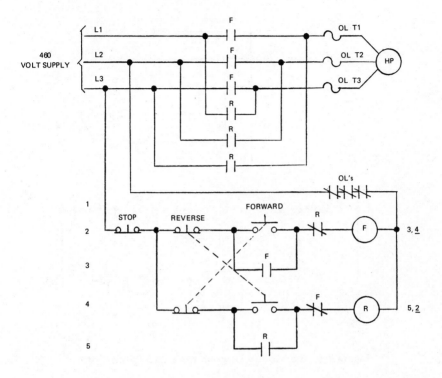

Figure 7-5. Elementary Diagram for a FVR Motor

Two-Speed Motors

To change the speed of the motor requires changing the effective number of poles. This is accomplished either by using a motor with two separate windings or a motor whose windings are taped so that they can be connected in two ways. Six power leads are required to change the windings. Figure 7-6 illustrates an elementary diagram for a two-speed motor with separate windings. The control portion of the elementary is similar to that of a reversing motor (Figure 7-5) except that six overloads are required.

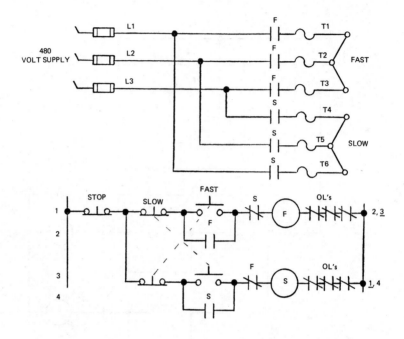

Figure 7-6. Elementary Diagram for a Two-Speed Motor

DESCRIPTION OF OPERATION
AND LOGIC DIAGRAMS

To describe how a process operates it is necessary to establish a logic diagram or description of operation. From this description the electrical schematic is designed.

SIM 7-1

From the description of operations, draw an electrical schematic.

The HVAC design is as follows:

(a)   When exhaust fan #1 is operated, damper EP valve is energized (electric to pneumatic).

(b)   PE (pneumatic to electric) switch prevents #2 fan from starting.

Answer

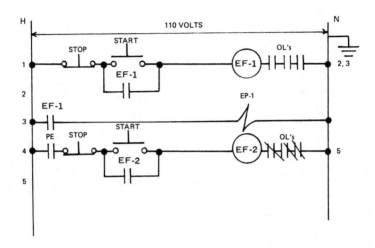

SIM 7-2

Draw elementaries for each scheme.

(a) When Level Switch LCH #1 in Tank "A" reaches "high" level, Pump #3 is started. Pump will run for ten minutes before it will turn off automatically.

(b) Pump #4 which feeds Tank "A" can be started manually and will automatically stop when high level occurs (LCHA).

(c) There are three valves which are used at a manifold station. When Valve "A" is energized Valve "C" can not be energized. When Valve "B" is energized, Valves "A" and "C" can not be energized. Once Valve "C" is energized, Valves "A" and "B" have no effect.

Answer

It should be noted that several electrical schematics may all be correct but look different. Another point is that several people may interpret the above descriptions differently. One possible solution is illustrated.

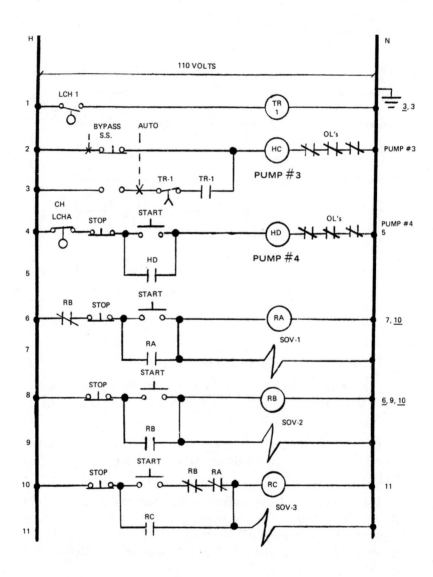

LOGIC DIAGRAMS AND MINI-COMPUTERS

Logic diagrams use symbols to describe the operation whereas the elementary diagram uses words. The logic diagram is an effective tool to convey process requirements since it is understood by all engineers. To describe a process, the logic diagram uses symbols for the three words "And," "Or," "Not." Since any logic can be conveyed with these three words it offers a simple, exact means to describe process requirements. Figure 7-7 illustrates symbols commonly used in logic. These symbols are combined to describe a process. It is possible to design a complete electrical control scheme by using a mini-computer and a logic diagram. A mini-computer is basically a packaged combination of memory units which can be easily programmed for any process. Unlike relays they are not hard-wired, so that changes in operation can be readily made by just reprogramming the unit. It is possible to use the input directly from the logic diagram to program the mini-computer without drawing the elementary diagram as an intermediate step.

Thus logic diagrams can save design time when used with new programmable devices. They are also useful in the development of conventional control schemes since they offer a visual tool which can be understood not only by the electrical engineer, but by the process and mechanical engineer as well. This means that prior to the start of the electrical elementaries and interconnecting drawings the process can be resolved. This in itself can minimize costly design changes.

HOW TO ANALYZE LOGIC DIAGRAMS

To analyze a logic diagram it is necessary to determine the various inputs required to actuate the logic gate. When time delays are incorporated into the design it is necessary to determine the "state" of the process at various time periods. One method to accomplish this is to place a "1" or a "0" after each logic symbol.

A "1" indicates a signal present and a "0" indicates the absence of a signal. Always try to establish the initial state of the circuit. The logic elements can be combined in any order to describe the electrical circuit.

Logic Functions	NEMA Symbol	Description
AND		A device which produces an output only when every input is present—represents contacts in series.
OR		A device which produces an output when one input (or more) is present—represents contacts in parallel.
NOT		A device which produces an output only when the input is absent — represents a normally closed contact.
ON DELAY TIMER		A device which produces an output following a definite time delay after its input is applied.
OFF DELAY TIMER		A device whose output is removed following a definite time delay after its input is removed.

Figure 7-7. Logic Symbols

SIM 7-3

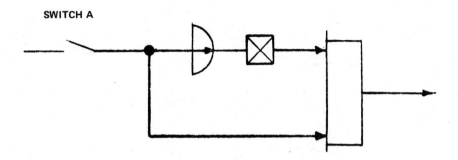

SWITCH A

What will happen when Switch "A" is closed?

Draw an elementary.

Answer

Step 1   Identify *initial* operating and final state of logic.

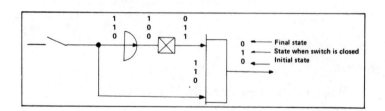

Comments—The above logic shows how various elements can be combined. Note that two signals from switch "A" are used. One goes through a "Time Delay" and "Not" gate and the other is an instantaneous signal fed directly into an "And" gate. Both signals are required to give a pulse after the switch is closed.

Step 2   Look at output — a pulse when equipment is activated.

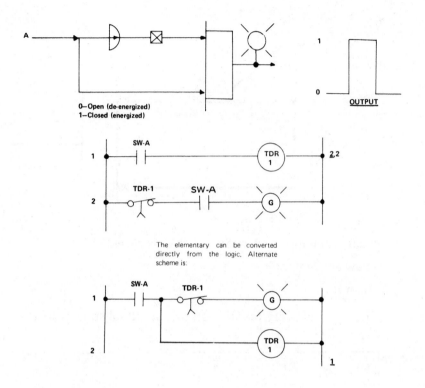

0—Open (de-energized)
1—Closed (energized)

The elementary can be converted directly from the logic. Alternate scheme is:

SIM 7-4

What will happen when Sw. 1 is closed?
Draw an elementary diagram.

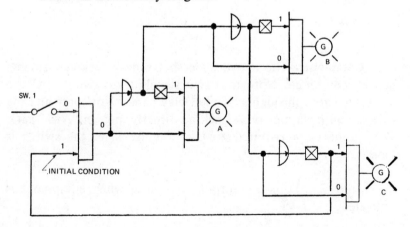

Answer

How to Analyze Logic

(a) Determine initial condition with switch open.

(b) By placing "1s" and "0s" on above diagram determine operation when switch is closed.

(c) Always check final state to see if logic resets.

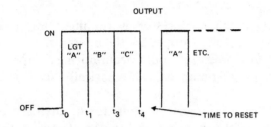

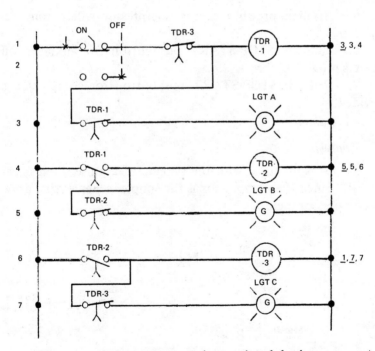

The above elementary and associated logic represent sequential pulses when the switch is closed.

SIM 7-5

From the description of operation, draw a logic diagram and an elementary.

1.   In order for the compressor to operate (C-1), the Oil Pump (P-2) should be running for 5 minutes. An interlock should be provided so that if an operator pressed the start button, he must keep the button depressed until the above is accomplished.

2.   After the compressor is running, the auxiliary oil pump will be manually stopped by the operator.

3.   If low pressure should occur during compressor operation, auxiliary oil pump will automatically start. (Stop same as 2.)

4.   When compressor is shut down, auxiliary oil pump will automatically come on and automatically stop after 5 minutes.

5.   If high pressure occurs compressor will automatically stop.

6.   Manual STOP-START pushbutton for auxiliary oil pump is provided.

7.   Manual STOP-START pushbutton for compressor is provided.

Answer

Analysis

Step 4 is identical with the scheme of a pulse when signal goes off. Since the compressor is the simpler scheme, first draw the compressor logic.

Compressor Logic

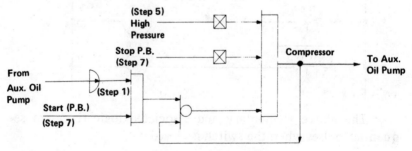

In the auxiliary oil pump scheme the motor can be manually started or started through low pressure. Since Step 4 also automatically stops the motor, this step must be drawn independently of the seal in contact.

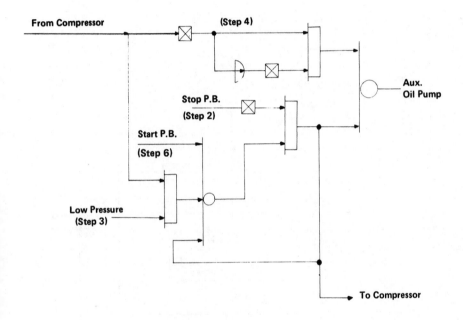

The two basic schemes can be combined, as illustrated on the following page.

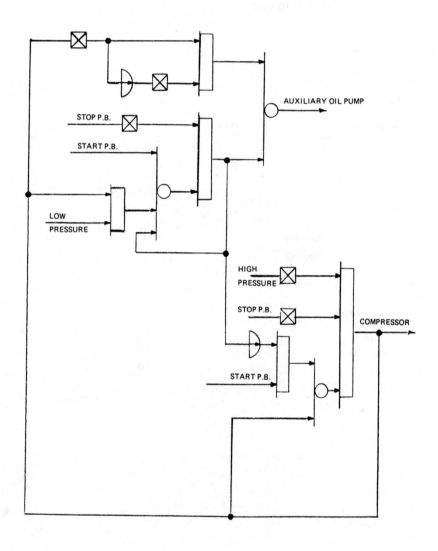

Development of Elementary

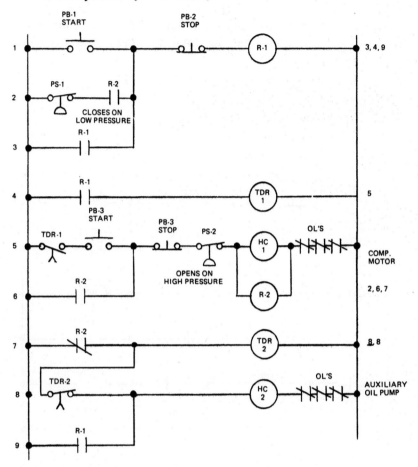

INTERCONNECTION DIAGRAMS

To show the electrician how to wire from an elementary diagram, a physical arrangement referred to as an interconnection diagram is drawn. Vendor's prints showing how the terminal blocks are arranged are used. Each wire on the elementary is assigned a wire number. All devices are connected internally to the terminal blocks for each panel. The electrician needs only to connect the terminals together with the control cable.

SIM 7-6

Draw an interconnection diagram for the elementary below. Wire numbers and terminal numbers have the same designation as indicated.

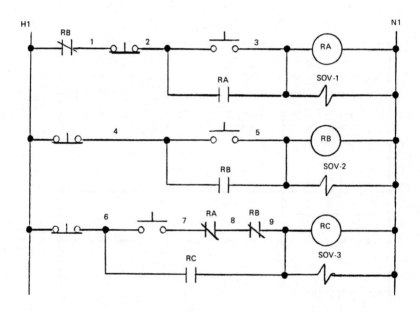

Pushbuttons are located on Pushbutton Panel No. 1. Relays are located on Relay Panel No. 1. Solenoid valves are located locally; 110 volts source is from M.C.C. No. 1.

Answer

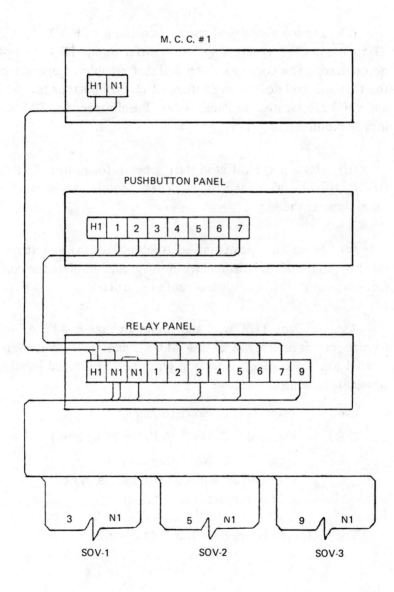

JOB SIMULATION–SUMMARY PROBLEM

JOB 6

(a)   Draw an electrical control scheme for CF-3, C-16 and TP-5. When CF-3 is reversing, the transfer pump TP-5 can not be operated. The conveyor C-16 will not be able to operate in the forward cycle of the centrifuge CF-3. Pushbuttons for C-16 and CF-3 are located on Panel No. 1. Pushbuttons for TP-5 are locally mounted.

(b)   Draw a typical stop-start scheme for motors CTP-6, HF-11, HF-11, BC-13 and SC-19. These motors can be started locally and at the M.C.C.

(c)   From the elementaries developed in (a) and (b) of this problem, and the elementary illustrated, complete the control portions of the conduit and cable schedule.

Assume Type THW wire and minimum size is #14. All interconnections are made at the M.C.C. and field devices are located near each other. Common wires will be jumped locally. Designate conduits as follows:

C-P1   From M.C.C. No. 1 to Panel No. 1.

C-R1   From M.C.C. No. 1 to Relay Panel No. 1.

C-L1   From M.C.C. No. 1 to local devices
           (exclude local pushbuttons since those
           cables were sized in Job 5).

The motor list is based on Job 1, Chapter 2.

Illustrated elementary [see (c)] for motors AG-1, FP-4, CT-9, UH-12 and RD-22.

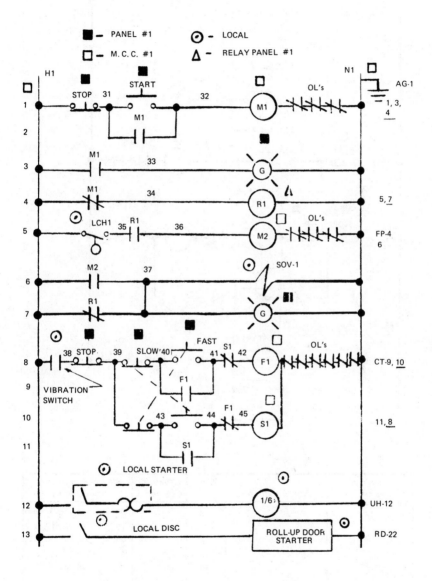

Analysis

(a) and (b)

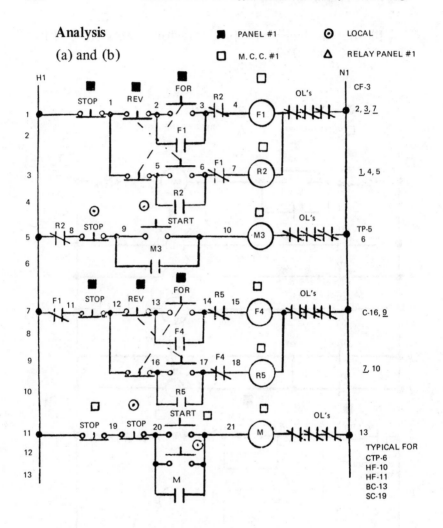

(c)

Conduit Designation	No. Conductors	Wire Size	Conduit Size	Wire Numbers in Each Conduit
C-P1	20	#14	1¼"	H1, N1, 2, 3, 5, 6, 11, 13, 14, 16, 17, 31, 32, 33, 37, 38, 40, 41, 43, 44
C-R1	6	#14	¾"	H1, N1, 34, 35, 36, 37
C-L1	5	#14	¾"	H1, N1, 35, 37, 38

SUMMARY

Many times it is difficult to obtain a description of operation from the process, project or mechanical engineer. In many plants the elementary diagram is the only document which describes how a process works. Usually only the electrical engineer understands the elementary diagram so it is difficult to insure that the elementary diagram meets the functional requirements. The verbal description which is the input to the elementary diagram can be easily misinterpreted. Since all disciplines understand the simple logic elements, logic diagrams can be initiated and agreed upon. The electrical engineer can then use the logic diagram as the basis for his design. By representing the circuit in terms of a logic diagram the electrical engineer can use Boolean algebra or other switching circuit techniques to simplify the elementary control. Also state-of-the-art techniques using solid-state logic elements or programmable controllers can be specified directly from the logic diagram.

The logic elements form a common language permitting the process, mechanical, project and electrical engineers to create a complex elementary control scheme. The individual who understands the fundamentals of the logic elements can easily apply this important tool in developing complex industrial and power elementary control schemes.

Chapter 8

APPLYING PROCESS
CONTROLLERS AND ELECTRONIC
INSTRUMENTATION

The basic requirement for developing electrical schematics is to understand the process. The logic diagram is of particular use in describing a process when the activating input is related to an on-off signal. Traditionally, the hardware used to accomplish this logic has been a combination of heavy-duty machine tool relays. A growing trend is the use of solid-state programmable controllers, especially in applications where the process steps repeat continuously. In addition, the current trend is to control process variables, i.e., temperature, pressure, flow level by electronic instrumentation loops. Thus it becomes increasingly important to understand how programmable controllers and instrumentation are applied.

PROGRAMMABLE CONTROLLER

The programmable controller developed out of the needs of the automotive industry. The industry required a control unit which could easily be changed in the plant, easily be maintained and repaired, highly reliable, small, capable to output data to a central data collection system, and competitive in cost to relay and solid-state panels. Out of these requirements developed the first series of programmable controllers. These units were initially designed to accept 115-volt ac inputs, to output 115 volts ac −2 ampere signals capable of actuating solenoid valves, motors, etc. The memory was capable of expansion to a maximum of 4000 words.

Today the programmable controller is far more flexible and reliable than the earlier generations and its use has had far-reaching implications to most industrial plants. Let's examine some of the features of most models.

The basic unit described in Figure 8-1 consists of:

- *Input module*
- *Memory*
- *Processor*
- *Output module*
- *Programming auxiliary*

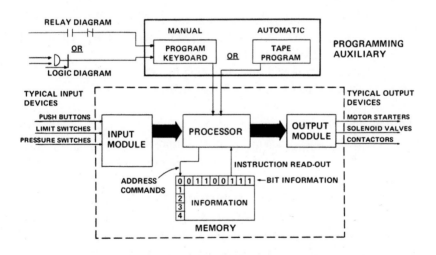

Figure 8-1. Components of a Programmable Controller

Input Module

The input module usually accepts ac or dc signals from remote devices such as pushbuttons and switches. These signals are then converted to dc levels, filtered and attenuated for use by the processor.

Processor

This module is the working portion of the programmable controller. All input and outputs are continuously monitored. The status of the input is compared against an established program and instructions are executed to the various outputs. The control function of the processor identifies the memory core locations to be addressed. An internal timing device determines the required sequence to fetch and execute instructions.

Memory

The information stored in the memory relates to how the input-output data should be processed. Information is usually stored on magnetic cores.

Output Module

The output module provides the means to command external machine devices. Output loads are usually energized through triac ac switches or reed contacts.

Many units have input and output lights on the unit which indicate the status of the remote devices.

Programming Auxiliary

The programming auxiliary enables the electrical consultant to communicate with the unit. Manufacturers use slightly different approaches to the interface problem but the two common typical methods for programming are:

Use of external keyboard. Keyboard may use logic symbols or standard elementary symbols. In either case, the logic or elementary is directly programmed into the unit. Refer to Figure 8-2.

Use of Storage Media: Some units offer keyboards with visual displays. The visual displays are useful when input data needs to be verified and when changes occur in the elementary. A program can be stored on a magnetic disk or magnetic tape. This allows the program to be developed in one location and transferred to the site. Additionally, this allows a backup copy of the program to be available if the system goes down.

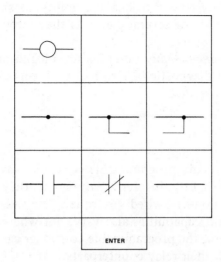

Figure 8-2. Typical Keyboard for Programmable Controller

Another feature of programming is in elementary simplification and simulation. If a tape or disk is used for programming, the information on the tape can be used in a standard elementary simplification program. Any redundancies in the logic can be quickly found. The tape or disk can also be used in conjunction with a simulation program. For any input, the corresponding output can be determined without the actual output being energized. To change sequence of operation, the memory is simply reprogrammed.

Capabilities

The programmable controller is capable of performing the function of hardwired solid-state systems or relay systems.

The basic functions are:

- *Logic Gates*—For a review of logic, see Chapter 7.
- *Timers*—Either on delay or off delay timers. Typical range .1 to 99.9 seconds.
- *Counters*—Used to count input status changes at a rate of 20 per second. Counter size 999 counts.
- *Latches*—Simulates electro-mechanical relay.

- *Shift Registers*—Provides the ability to simultaneously remember the state of several pieces as they move through the manufacturing process.
- *Maintenance*—Many units feature plug-in modules and a preprogrammed diagnostic tape so that problems can be quickly identified and corrected.

Applications

There are many programmable controllers available. Each unit may have slightly different features but they all offer the versatility of a non-hardwired controller. The size of the input-output and logic capabilities also vary, but where repeat operations are present, the programmable controller can economically compete with their relay counterparts.

In the case where the units are economically competitive, the programmable controller should seriously be considered since it offers:

- Complete flexibility to modify or expand controls when production needs change.

- A highly reliable solid state control which minimizes downtime.

- The possible reduction of engineering design time (no need to check back of panel wiring, etc.).

- A system compatible with computer simplification and simulation techniques.

- Fast diagnostics and maintenance.

INSTRUMENTATION

Modern Process Control hinges around electronic instrumentation. For years pneumatic instrumentation dominated the market, but today most control loops have been replaced by their electronic counterparts.

Closed Loop

To control a process requires a closed loop system. In the example of controlling level, if the level is too high the level is reduced. Figure 8-3 shows a typical closed loop system.

Where:

C   is the controlled variable

R   is the reference or set point

E   is the error or deviation

M   is the variable manipulated by the controller.

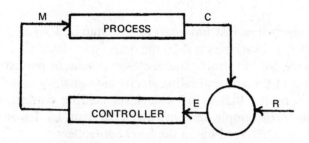

Figure 8-3. Typical Closed Loop System

Figure 8-4 shows a typical process control diagram illustrating level control.

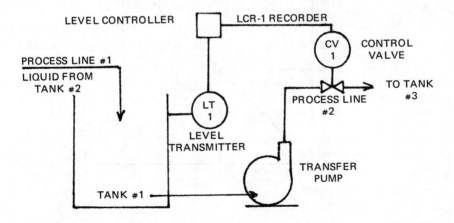

Figure 8-4. Typical Level Control Process

In this process level transmitter (LT-1) sends the electronic signal to level controller recorder (LCR-1). If the level is too low, control valve CV-1 is throttled. On the other hand, if the level is too high CV-1 is opened. Thus, the process is controlled. This example illustrates the fundamental elements of a control loop, namely:

Input (transmitter)
Controller
Output (control valve).

Signal

In electronic instrumentation the standard signal used to convey process variables is 4-20 ma dc.

Figure 8-5 illustrates the process variable in percentage as a function of the corresponding electronic signal.

This means that a full-scale level reading corresponds to 100%. In this example an 8 ma dc signal from LT-1 would correspond to a 25% reading on the level controller.

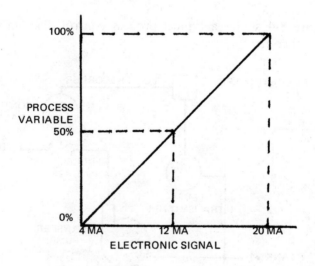

Figure 8-5. Process Variable vs. Electonic Signal

Control Loop

The basic control circuitry for electronic instrumentation is illustrated in Figure 8-6. This figure illustrates a typical series circuit for the process of Figure 8-4. Each receiving element contains an input impedance.

In a series network, the impedance of each element must be carefully matched.

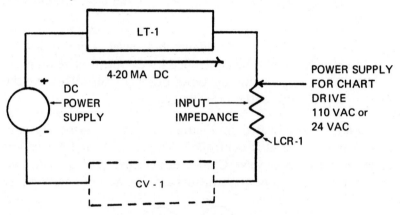

Figure 8-6. Control Loop

DIRECT DIGITAL CONTROL

Process controllers are beginning to be used to make not only logic decisions but also control loop decisions. Figure 8-7 shows a process controller configuration to accomplish the process of Figure 8-4. Such a configuration is often called direct digital control (DDC) since the control action is determined by the program within the controller. DDC offers an easy way of changing control strategies without rewiring or recalibrating. Additionally, DDC requires much less calibration than do pneumatic and electronic controllers.

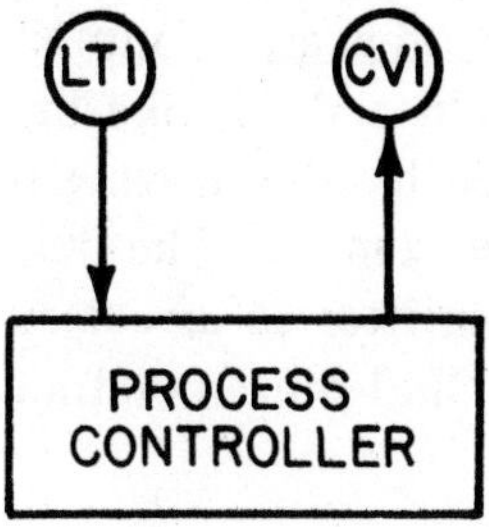

Figure 8-7.

Direct Digital Control

Types of Control Instrumentation

The several common types of control variables are:

- *Level*
- *Flow*
- *Pressure*
- *Temperature*

Level: As indicated in Figure 8-4, level is controlled by CV-1. Level could also be controlled by a control valve on process line No. 1. Two common types of level transmitters are the float type and differential pressure type.

Flow: As indicated in Figure 8-8 flow is essentially controlled the way level is controlled The signal from FT-1 is sent to FC-1 and control valve CV-1 is either opened or closed. Two common types of flow transmitters are the magnetic flow meter and orifice meter type.

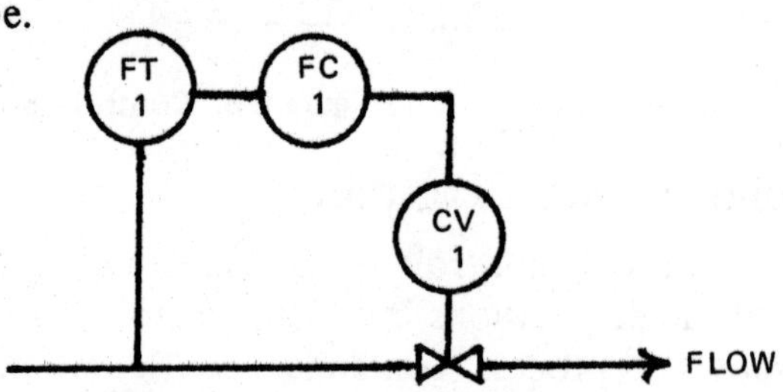

Figure 8-8.
Flow Control

Magmeter Type: The magnetic flow meter is the more expensive of the two. The magnetic flow meter uses the same principle of operation as a tachometer or generator. As indicated in Figure 8-9 the fluid flowing acts as the conductor while the pipe is located in the magnetic field caused by the field coils. The electrodes are mounted in a plane at right angles to the magnetic field and act like brushes of a generator. Thus the voltage induced by the moving fluid is brought out by leads for external measurement. The induced voltage can be converted by the transmitter to direct current. For magmeters to work properly, the fluid must conduct electricity and the tube must be liquid full.

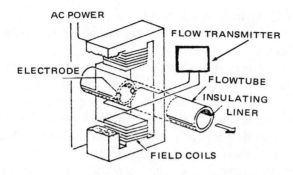

Figure 8-9. Magnetic Flow Meter

Orifice Meter Type: As liquid flows through a restriction in a pipe orifice plate, a change in pressure occurs as illustrated in Figure 8-10. The flow rate is calculated by measuring the differential pressure across the orifice plate.

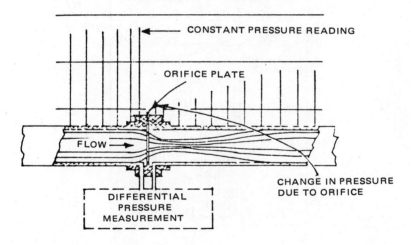

Figure 8-10. Orifice Flow Meter

Pressure: Pressure control is used to maintain a specified pressure or is used as an indirect measurement of level or flow, as indicated previously.

Temperature: The thermocouple is one of the most frequently used methods for measuring temperatures between 500 and 1500°C. Strictly speaking the thermocouple and its associated control do not fall into the typical electronic instrumentation category.

The operation of a thermocouple is based on the principle that an electromotive force (emf) is developed when two different metals come in contact. The emf developed is dependent on the metals involved and the *temperature* of the junctions. The emf signal developed is expressed in millivolts. Due to the nature of the thermocouple, wire splices should be avoided. Two types of measuring circuits are used in conjunction with thermocouples.

Figure 8-11 illustrates a typical thermocouple circuit using a Galvanometer.

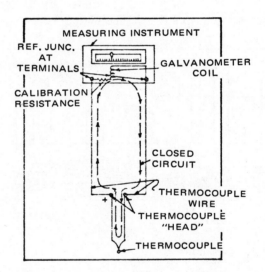

Figure 8-11. Basic Galvanometer Thermocouple Circuit

Figure 8-12 illustrates a typical thermocouple circuit using a potentiometer.

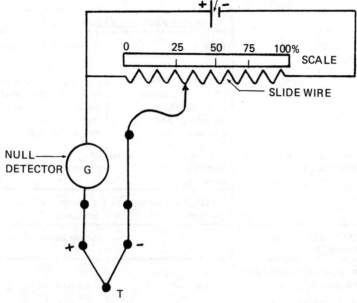

Figure 8-12. Basic Potentiometer Thermocouple Circuit

The RTD (resistance temperature device) is becoming most common in measuring temperatures. The RTD resistance varies with temperature and consequently when it is installed in a circuit the resulting voltage drop is proportional to the temperature.

WIRING METHODS

The type of cable, and installation of instrument signals is based on the reduction of noise. In general instrument cables are routed away from noise sources such as power cables, motors, generators, etc. Twisted control cables are used to reduce magnetic noise pickup from a nearby noise source.

Table 8-1 summarizes instrumentation wiring methods.

Table 8-1. Instrumentation Wiring Methods

	RECEIVING ELEMENT DESCRIPTION	
TRANSMITTER TYPE	Torque type	Bridge or potentiometer type
	current converted directly to a torque to move a chart recorder or pointer, i.e., pyrometer, Galvanometer	input signal is compared with a standard voltage and amplified to drive a chart or recorder to a balanced or null position
THERMOCOUPLE	twisted pair nonshielded	twisted pair shielded
	NOTE: lead wire must be the same material as the thermocouple, i.e., iron constantan	
MAGNETIC FLOW METERS		twisted pair shielded
PNEUMATIC CURRENT TRANSDUCERS, DIFFERENTIAL PRESSURE FLOW METERS	twisted pair nonshielded	*twisted pair nonshielded *NOTE: for a high noise environment use twisted pair shielded.

Chapter 9

PROTECTIVE RELAYING
FOR POWER
DISTRIBUTION SYSTEMS

Due to possible equipment failure or human error, it is necessary to provide protection devices. These devices minimize system damage and limit the extent and duration of service interruption when failure occurs. The main goal of protection coordination is to isolate the affected portion of the system quickly while at the same time maintaining normal service for the remainder of the system. In other words, the electrical system must provide protection and selectivity to insure that the fault is minimized while other parts of the system not directly involved are held in until other protective devices clear the trouble. Protective devices such as fuses and circuit breakers have time current characteristics which determine the time it takes to clear the fault. In the case of circuit breakers it is possible to adjust the characteristics while fuse characteristics are non-adjustable.

Protective relays are another way of achieving selective coordination and are required to operate power breakers above 600 volts. By definition a protective relay is a device which when energized by suitable currents, voltages, or both, responds to the magnitude and relationships of these signals to indicate or isolate an abnormal operating condition. These relays have adjustable settings and can be used to actuate the opening of circuit breakers under various fault conditions.

In Chapter 4, the concept of the one-line diagram was introduced. In order to complete the power distribution system it is necessary to show on the one-line diagram the protective relaying required and the breakers affected. To properly set the protective devices it is necessary to know the fault currents which occur at various portions of the system. This chapter will illustrate typical applications of protective devices.

THE OVERCURRENT RELAY (Device 51)

Every protective relay has an associated number. Some of the standard designations are listed in Table 9-1. The overcurrent relay is designated as Device 51. This relay is used for overcurrent protection and is current sensitive. The one-line diagram utilizing the overcurrent relay is shown in Figure 9-1.

Table 9-1. Protective Device Numbering and Functions

DEVICE NUMBER	DEFINITION AND FUNCTION
1	**master element** is the initiating device, such as a control switch, voltage relay, float switch, etc., which serves either directly, or through such permissive devices as protective and time-delay relays to place an equipment in or out of operation.
2	**time-delay starting, or closing, relay** is a device which functions to give a desired amount of time delay before or after any point or operation in a switching sequence or protective relay system, except as specifically provided by device functions 62 and 79 described later.
3	**checking or interlocking relay** is a device which operates in response to the position of a number of other devices, or to a number of predetermined conditions in an equipment to allow an operating sequence to proceed, to stop, or to provide a check of the position of these devices or of these conditions for any purpose.
4	**master contactor** is a device, generally controlled by device No. 1 or equivalent, and the necessary permis-

Table 9-1. Protective Device Numbering and Functions
(continued)

DEVICE NUMBER	DEFINITION AND FUNCTION
4 (con't.)	sive and protective devices, which serves to make and break the necessary control circuits to place an equipment into operation under the desired conditions and to take it out of operation under other or abnormal conditions.
5	**stopping device** functions to place and hold an equipment out of operation.
6	**starting circuit breaker** is a device whose principal function is to connect a machine to its source of starting voltage.
7	**anode circuit breaker** is one used in the anode circuits of a power rectifier for the primary purpose of interrupting the rectifier circuit if an arc back should occur.
8	**control power disconnecting device** is a disconnecting device—such as a knife switch, circuit breaker or pull-out fuse block—used for the purpose of connecting and disconnecting, respectively, the source of control power to and from the control bus or equipment. **note:** Control power is considered to include auxiliary power which supplies such apparatus as small motors and heaters.
9	**reversing device** is used for the purpose of reversing a machine field or for performing any other reversing functions.
10	**unit sequence switch** is used to change the sequence in which units may be placed in and out of service in multiple-unit equipments.
11	Reserved for future application.
12	**over-speed device** is usually a direct-connected speed switch which functions on machine overspeed.
13	**synchronous-speed device,** such as a centrifugal speed switch, a slip-frequency relay, a voltage relay, an undercurrent relay or any type of device, operates at approximately synchronous speed of a machine.
14	**under-speed device** functions when the speed of a machine falls below a predetermined value.

/more/

Table 9-1. Protective Device Numbering and Functions (continued)

DEVICE NUMBER	DEFINITION AND FUNCTION
15	**speed or frequency, matching device** functions to match and hold the speed or the frequency of a machine or of a system equal to, or approximately equal to, that of another machine, source or system.
16	Reserved for future application.
17	**shunting, or discharge, switch** serves to open or to close a shunting circuit around any piece of apparatus (except a resistor), such as a machine field, a machine armature, a capacitor or a reactor. **note:** This excludes devices which perform such shunting operations as may be necessary in the process of starting a machine by devices 6 or 42, or their equivalent, and also excludes device 73 function which serves for the switching of resistors.
18	**accelerating or decelerating device** is used to close or cause the closing of circuits which are used to increase or to decrease the speed of a machine.
19	**starting-to-running transition contactor** is a device which operates to initiate or cause the automatic transfer of a machine from the starting to the running power connection.
20	**electrically operated valve** is a solenoid- or motor-operated valve which is used in a vacuum, air, gas, oil, water, or similar, lines. **note:** The function of the valve may be indicated by the insertion of descriptive words such as "Brake" or "Pressure Reducing" in the function name, such as "Electrically Operated **Brake** Valve."
21	**distance relay** is a device which functions when the circuit admittance, impedance or reactance increases or decreases beyond predetermined limits.
22	**equalizer circuit breaker** is a breaker which serves to control or to make and break the equalizer or the current-balancing connections for a machine field, or for regulating equipment, in a multiple-unit installation.
23	**temperature control device** functions to raise or to lower the temperature of a machine or other apparatus,

/more/

Table 9-1. Protective Device Numbering and Functions
(continued)

DEVICE NUMBER	DEFINITION AND FUNCTION
23 (con't.)	or of any medium, when its temperature falls below, or rises above, a predetermined value. **note:** An example is a thermostat which switches on a space heater in a switchgear assembly when the temperature falls to a desired value as distinguished from a device which is used to provide automatic temperature regulation between close limits and would be designated as 90T.
24	Reserved for future application.
25	**synchronizing, or synchronism-check, device** operates when two a-c circuits are within the desired limits of frequency, phase angle or voltage, to permit or to cause the paralleling of these two circuits.
26	**apparatus thermal device** functions when the temperature of the shunt field or the armortisseur winding of a machine, or that of a load limiting or load shifting resistor or of a liquid or other medium exceeds a predetermined value; or if the temperature of the protected apparatus, such as a power rectifier, or of any medium decreases below a predetermined value.
27	**undervoltage relay** is a device which functions on a given value of undervoltage.
28	Reserved for future application.
29	**isolating contactor** is used expressly for disconnecting one circuit from another for the purposes of emergency operation, maintenance, or test.
30	**annunciator relay** is a nonautomatically reset device which gives a number of separate visual indications upon the functioning of protective devices, and which may also be arranged to perform a lockout function.
31	**separate excitation device** connects a circuit such as the shunt field of a synchronous converter to a source of separate excitation during the starting sequence; or one which energizes the excitation and ignition circuits of a power rectifier.
32	**directional power relay** is one which functions on a desired value of power flow in a given direction, or upon reverse power resulting from arc back in the anode or cathode circuits of a power rectifier.

/more/

Table 9-1. Protective Device Numbering and Functions
(continued)

DEVICE NUMBER	DEFINITION AND FUNCTION
33	**position switch** makes or breaks contact when the main device or piece of apparatus, which has no device function number, reaches a given position.
34	**motor-operated sequence switch** is a multi-contact switch which fixes the operating sequence of the major devices during starting and stopping, or during other sequential switching operations.
35	**brush-operating, or slip-ring short-circuiting, device** is used for raising, lowering, or shifting the brushes of a machine, or for short-circuiting its slip rings, or for engaging or disengaging the contacts of a mechanical rectifier.
36	**polarity device** operates or permits the operation of another device on a predetermined polarity only.
37	**undercurrent or underpower relay** is a device which functions when the current or power flow decreases below a predetermined value.
38	**bearing protective device** is one which functions on excessive bearing temperature, or on other abnormal mechanical conditions, such as undue wear, which may eventually result in excessive bearing temperature.
39	Reserved for future application.
40	**field relay** is a device that functions on a given or abnormally low value or failure of machine field current, or on an excessive value of the reactive component of armature current in an a-c machine indicating abnormally low field excitation.
41	**field circuit breaker** is a device which functions to apply, or to remove, the field excitation of a machine.
42	**running circuit breaker** is a device whose principal function is to connect a machine to its source of running voltage after having been brought up to the desired speed on the starting connection.
43	**manual transfer or selector device** transfers the control circuits so as to modify the plan of operation of the switching equipment or of some of the devices.

/more/

Table 9-1. Protective Device Numbering and Functions
(continued)

DEVICE NUMBER	DEFINITION AND FUNCTION
44	**unit sequence starting relay** is a device which functions to start the next available unit in a multiple-unit equipment on the failure or on the non-availability of the normally preceding unit.
45	Reserved for future application.
46	**reverse-phase, or phase-balance, current relay** is a device which functions when the polyphase currents are of reverse-phase sequence, or when the polyphase currents are unbalanced or contain negative phase-sequence components above a given amount.
47	**phase-sequence voltage relay** is a device which functions upon a predetermined value of polyphase voltage in the desired phase sequence.
48	**incomplete sequence relay** is a device which returns the equipment to the normal, or off, position and locks it out if the normal starting, operating or stopping sequence is not properly completed within a predetermined time.
49	**machine, or transformer, thermal relay** is a device which functions when the temperature of an a-c machine armature, or of the armature or other load carrying winding or element of a d-c machine, or converter or power rectifier or power transformer (including a power rectifier transformer) exceeds a predetermined value.
50	**instantaneous overcurrent, or rate-of-rise relay** is a device which functions instantaneously on an excessive value of current, or on an excessive rate of current rise, thus indicating a fault in the apparatus or circuit being protected.
51	**a-c time overcurrent relay** is a device with either a definite or inverse time characteristic which functions when the current in an a-c circuit exceeds a predetermined value.
52	**a-c circuit breaker** is a device which is used to close and interrupt an a-c power circuit under normal conditions or to interrupt this circuit under fault or emergency conditions.

/more/

**Table 9-1. Protective Device Numbering and Functions
(continued)**

DEVICE NUMBER	DEFINITION AND FUNCTION
53	**exciter or d-c generator relay** is a device which forces the d-c machine field excitation to build up during starting or which functions when the machine voltage has built up to a given value.
54	**high-speed d-c circuit breaker** is a circuit breaker which starts to reduce the current in the main circuit in 0.01 second or less, after the occurrence of the d-c overcurrent or the excessive rate of current rise.
55	**power factor relay** is a device which operates when the power factor in an a-c circuit becomes above or below a predetermined value.
56	**field application relay** is a device which automatically controls the application of the field excitation to an a-c motor at some predetermined point in the slip cycle.
57	**short-circuiting or grounding device** is a power or stored energy operated device which functions to short-circuit or to ground a circuit in response to automatic or manual means.
58	**power rectifier misfire relay** is a device which functions if one or more of the power rectifier anodes fails to fire.
59	**overvoltage relay** is a device which functions on a given value of overvoltage.
60	**voltage balance relay** is a device which operates on a given difference in voltage between two circuits.
61	**current balance relay** is a device which operates on a given difference in current input or output of two circuits.
62	**time-delay stopping, or opening, relay** is a time-delay device which serves in conjunction with the device which initiates the shutdown, stopping or opening operation in an automatic sequence.
63	**liquid or gas pressure, level, or flow relay** is a device which operates on given values of liquid or gas pressure, flow or level, or on a given rate of change of these values.

/more/

Table 9-1. Protective Device Numbering and Functions
(continued)

DEVICE NUMBER	DEFINITION AND FUNCTION
64	**ground protective relay** is a device which functions on failure of the insulation of a machine, transformer or of other apparatus to ground, or on flashover of a d-c machine to ground. **note:** This function is assigned only to a relay which detects the flow of current from the frame of a machine or enclosing case or structure of a piece of apparatus to ground, or detects a ground on a normally ungrounded winding or circuit. It is not applied to a device connected in the secondary circuit or secondary neutral of a current transformer, or current transformers, connected in the power circuit of a normally grounded system.
65	**governor** is the equipment which controls the gate or valve opening of a prime mover.
66	**notching, or jogging, device** functions to allow only a specified number of operations of a given device, or equipment, or a specified number of successive operations within a given time of each other. It also functions to energize a circuit periodically, or which is used to permit intermittent acceleration or jogging of a machine at low speeds for mechanical positioning.
67	**a-c directional overcurrent relay** is a device which functions on a desired value of a-c overcurrent flowing in a predetermined direction.
68	**blocking relay** is a device which initiates a pilot signal for blocking of tripping on external faults in a transmission line or in other apparatus under predetermined conditions, or co-operates with other devices to block tripping or to block reclosing on an out-of-step condition or on power swings.
69	**permissive control device** is generally a two-position, manually operated switch which in one position permits the closing of a circuit breaker, or the placing of an equipment into operation, and in the other position prevents the circuit breaker or the equipment from being operated.
70	**electrically operated rheostat** is a rheostat which is used to vary the resistance of a circuit in response to some means of electrical control.
71	Reserved for future application. */more/*

Table 9-1. Protective Device Numbering and Functions
(continued)

DEVICE NUMBER	DEFINITION AND FUNCTION
72	**d-c circuit breaker** is used to close and interrupt a d-c power circuit under normal conditions or to interrupt this circuit under fault or emergency conditions.
73	**load-resistor contactor** is used to shunt or insert a step of load limiting, shifting, or indicating resistance in a power circuit, or to switch a space heater in circuit, or to switch a light, or regenerative, load resistor of a power rectifier or other machine in and out of circuit.
74	**alarm relay** is a device other than an annunciator, as covered under device No. 30, which is used to operate, or to operate in connection with, a visual or audible alarm.
75	**position changing mechanism** is the mechanism which is used for moving a removable circuit breaker unit to and from the connected, disconnected, and test positions.
76	**d-c overcurrent relay** is a device which functions when the current in a d-c circuit exceeds a given value.
77	**pulse transmitter** is used to generate and transmit pulses over a telemetering or pilot-wire circuit to the remote indicating or receiving device.
78	**phase angle measuring, or out-of-step protective relay** is a device which functions at a predetermined phase angle between two voltages or between two currents or between voltage and current.
79	**a-c reclosing relay** is a device which controls the automatic reclosing and locking out of an a-c circuit interrupter.
80	Reserved for future application.
81	**frequency relay** is a device which functions on a predetermined value of frequency—either under or over or on normal system frequency—or rate of change of frequency.
82	**d-c reclosing relay** is a device which controls the automatic closing and reclosing of a d-c circuit interrupter, generally in response to load circuit conditions.
83	**automatic selective control, or transfer, relay** is a device which operates to select automatically between */more/*

Table 9-1. Protective Device Numbering and Functions
(continued)

DEVICE NUMBER	DEFINITION AND FUNCTION
83 (con't.)	certain sources or conditions in an equipment, or performs a transfer operation automatically.
84	operating mechanism is the complete electrical mechanism or servo-mechanism, including the operating motor, solenoids, position switches, etc., for a tap changer, induction regulator or any piece of apparatus which has no device function number.
85	carrier, or pilot-wire, receiver relay is a device which is operated or restrained by a signal used in connection with carrier-current or d-c pilot-wire fault directional relaying.
86	locking-out relay is an electrically operated hand or electrically reset device which functions to shut down and hold an equipment out of service on the occurrence of abnormal conditions.
87	differential protective relay is a protective device which functions on a percentage or phase angle or other quantitative difference of two currents or of some other electrical quantities.
88	auxiliary motor, or motor generator is one used for operating auxiliary equipment such as pumps, blowers, exciters, rotating magnetic amplifiers, etc.
89	line switch is used as a disconnecting or isolating switch in an a-c or d-c power circuit, when this device is electrically operated or has electrical accessories, such as an auxiliary switch, magnetic lock, etc.
90	regulating device functions to regulate a quantity, or quantities, such as voltage, current, power, speed, frequency, temperature, and load, at a certain value or between certain limits for machines, tie lines or other apparatus.
91	voltage directional relay is a device which operates when the voltage across an open circuit breaker or contactor exceeds a given value in a given direction.
92	voltage and power directional relay is a device which permits or causes the connection of two circuits when the voltage difference between them exceeds a given value in a predetermined direction and causes these

/more/

Table 9-1. Protective Device Numbering and Functions
(concluded)

DEVICE NUMBER	DEFINITION AND FUNCTION
92 (con't.)	two circuits to be disconnected from each other when the power flowing between them exceeds a given value in the opposite direction.
93	field changing contactor functions to increase or decrease in one step the value of field excitation on a machine.
94	tripping, or trip-free, relay is a device which functions to trip a circuit breaker, contactor, or equipment, or to permit immediate tripping by other devices; or to prevent immediate reclosure of a circuit interrupter, in case it should open automatically even though its closing circuit is maintained closed.
95 96 97 98 99	Used only for specific applications on individual installations where none of the assigned numbered functions from 1 to 94 is suitable.
notes:	[1] A similar series of numbers, starting with 201 instead of 1, shall be used for those device functions in a machine, feeder or other equipment when these are controlled directly from the supervisory system. Typical examples of such device functions are 201, 205, and 294.
	[2] A suffix X, Y, or Z denotes an auxiliary relay.
	[3] TC refers to trip coil.
	[4] CS refers to control switch.
	[5] N, G refers to neutral and ground respectively.

Overcurrent relays are available with inverse, very inverse and extremely inverse time current characteristics. The very inverse time current characteristic is the frequent choice when detailed system information is not available. The very inverse characteristic is most likely to provide optimum circuit protection and selectivity with other system protection devices. When

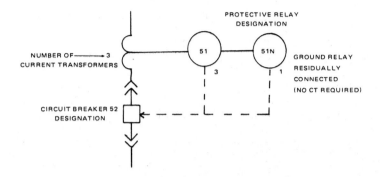

Figure 9-1. One-Line Diagram Showing Protective Relays

coordinating with fuses it may be necessary to use the extremely inverse characteristic.

Figure 9-2 illustrates the characteristics of electromechanical overcurrent protective relays. The overcurrent relays are current sensitive and require a seal-in contact to keep them energized after activation.

A common way in which power is supplied to a breaker for tripping purposes is through a d-c source using 125- or 250-volt station battery.

Figure 9-3 shows a typical electromechanical overcurrent relay schematic. Relay 51X is used as the seal-in relay which holds in the circuit until it is reset. For a three-phase circuit three of these relays are required.

Figure 9-4 illustrates a typical tripping circuit. Notice that protective relay contacts are connected in parallel so that any one will trip the breaker under fault conditions. The control switch (CS) can be manually used to trip the breaker. In order to simplify the protective relay elementary, many times a single contact is used to represent the three. Figure 9-5 presents a simplified protective relay elementary. To close the breaker, frequently just a manual close switch is used. A typical elementary used to close the circuit breaker is illustrated in Figure 9-6.

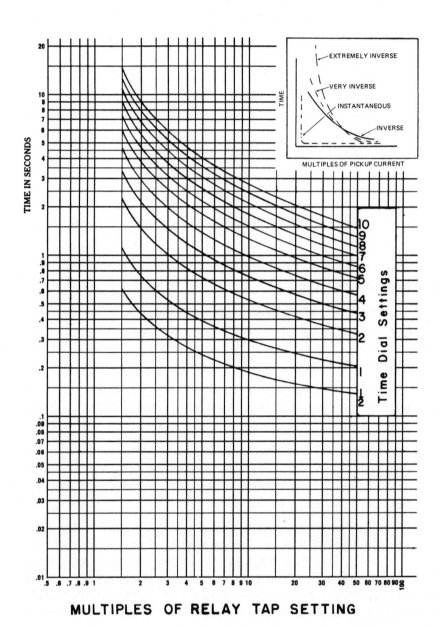

MULTIPLES OF RELAY TAP SETTING

Figure 9-2. Characteristics of Overcurrent Relays

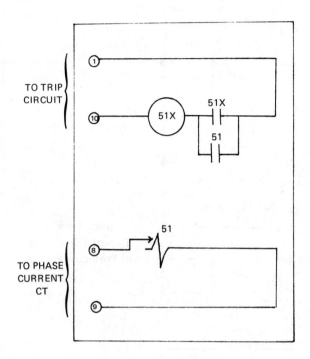

Figure 9-3. Typical Protective Relay (51) Schematic

**The Instantaneous
Overcurrent Relay
(Device 50)**

The instantaneous relay is usually combined with the over-current relay. The instantaneous attachment can be used or it can be disconnected from service if it is not required. Figure 9-7 illustrates a typical one-line diagram and schematic utilizing the 50, 51 and 51N protective relays.

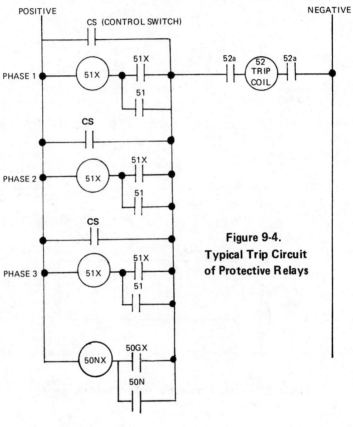

Figure 9-4.
Typical Trip Circuit
of Protective Relays

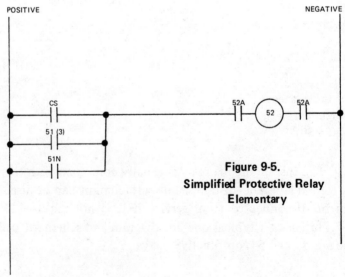

Figure 9-5.
Simplified Protective Relay
Elementary

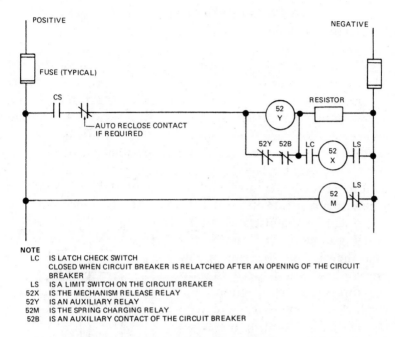

NOTE
LC   IS LATCH CHECK SWITCH
CLOSED WHEN CIRCUIT BREAKER IS RELATCHED AFTER AN OPENING OF THE CIRCUIT BREAKER
LS   IS A LIMIT SWITCH ON THE CIRCUIT BREAKER
52X   IS THE MECHANISM RELEASE RELAY
52Y   IS AN AUXILIARY RELAY
52M   IS THE SPRING CHARGING RELAY
52B   IS AN AUXILIARY CONTACT OF THE CIRCUIT BREAKER

Figure 9-6. Typical Close Elementary of a Circuit Breaker

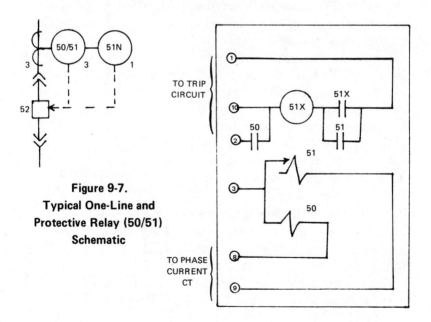

Figure 9-7.
Typical One-Line and
Protective Relay (50/51)
Schematic

THE GROUND OVERCURRENT RELAY
(Devices 50N and 50G)

Ground-fault protection has been required since the 1971 NEC. Ground-fault protection saves lives by minimizing damage to circuit conductors and other equipment and safeguarding persons who may simultaneously make contact with electrical equipment and a low resistance path to ground. Ground overcurrent protection can be provided either by overcurrent or instantaneous overcurrent relays. There are three common connections used for ground overcurrent relays; namely, residual connection, ground sensor connection, and the neutral CT connection.

Figure 9-8 illustrates the **residual grounding** scheme. Notice that the one-line diagram of Figure 9-7 has been detailed to show the three-phase connections. The residual relaying scheme detects ground-fault current by measuring the current remaining in the secondary of the three-phase of the circuit as transformed by the current transformers. Care must be taken to set the pick-up of the relay above the level anticipated by unbalanced single-phase loads. Due to the possible unbalances caused by unequal current transformer saturation on phase faults and transformer energizing inrush currents, the instantaneous overcurrent relay is seldom used.

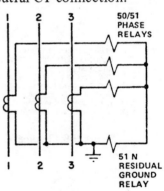

Figure 9-8.
Residual Grounding Scheme

The **ground sensing** scheme is illustrated in Figure 9-9. This scheme uses zero sequence current transformers to detect on ground faults the unbalances in the magnetic flux surrounding the three-

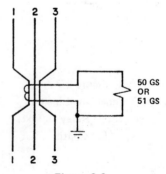

Figure 9-9.
Ground Sensor Scheme

phase conductors. Zero sequence current transformers detect when the vectorial summation of the currents is not zero.

The instantaneous or overcurrent relay can be used with this scheme. The installation of the zero-sequence window current transformer should not enclose the equipment ground conductor or the conductor shielding. With the ground sensing scheme it is possible to detect and clear system faults as small as 15 amperes.

The **neutral grounding** scheme illustrated in Figure 9-10 is used commonly with resistively grounded transformers. In this scheme the ground-fault current is sensed by the current transformer in the resistively grounded neutral conductor.

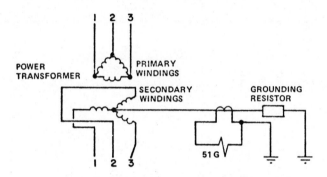

Figure 9-10. Neutral Grounding Relay Scheme

PARTIAL DIFFERENTIAL OR
SUMMATION RELAYING

This protective relaying scheme is commonly used to detect and isolate faults without affecting other portions of the system. Figure 9-11 illustrates a typical partial differential relaying scheme. In this scheme the tie breakers are nominally closed. If a fault occurs on BUS G, breakers C and D should trip leaving breaker A unaffected. Likewise a fault on BUS F should trip breakers A and C and leave breaker D unaffected. One way of accomplishing this is to connect the current transformers of

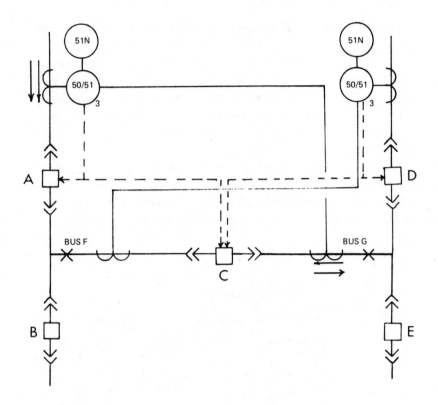

Figure 9-11. Partial Differential Protective Relay Scheme

protective relays such that they will only pick up when the fault currents through the pair of current transformers flow in opposite directions. For example, a fault on BUS F will cause fault currents flowing *to* the fault. The current transformers associated with protective relays for breaker A will sense currents in opposite directions thus activating these relays. On the other hand, a fault on BUS G will cause currents to flow in the same direction through the current transformers, thus these relays will not operate under this condition.

DIFFERENTIAL PROTECTIVE RELAY
(Device 87)

The differential protective relay is used for protecting a-c rotating machinery, generators and transformers. This relay operates on the difference between two currents. A typical application is illustrated in Figure 9-12. This figure shows the differential principle applied to a single-phase winding of electrical equipment such as a generator. In this application a current balance relay is used to provide what is called "percentage differential" relaying. The current transformers are connected to the equipment to be protected.

The current from each transformer flows through a restraining coil. The purpose of the restraining coil is to prevent undesired relay operation as a result of a mismatch in current transformers. When a fault does occur, the operating relay sees a percentage increase in current and the relay operates.

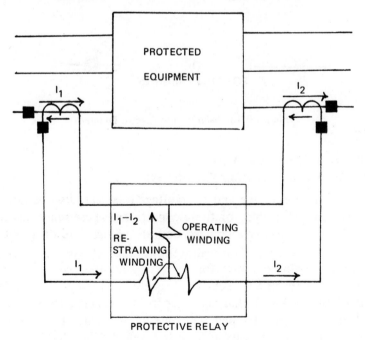

Figure 9-12. Differential Protective Relay Scheme

Notice on Figure 9-12 the introduction of polarity marks on the current transformers. Polarity identification marks are as follows:

  • Current flows into the polarity mark for primary connections. Current flows out of the polarity mark for secondary connections.

  • Voltage drop from polarity to nonpolarity for primary and secondary connections.

See Figure 9-13 for an illustration of polarity marks.

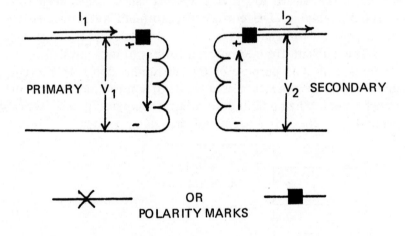

Figure 9-13. Polarity and Circuit Diagram

UNDERVOLTAGE RELAY (Device 27)
AND OVERVOLTAGE RELAY (Device 59)

The undervoltage and overvoltage relays are used wherever protection is required for these conditions. These relays usually operate continuously energized and are adjusted to drop out at any voltage within their calibration range. Figure 9-14 illustrates a one-line diagram for undervoltage and overvoltage relays. In this particular scheme, breaker *A* is tripped by undervoltage relay. An auxiliary contact from breaker *A* trips breaker *B* after breaker *A* is tripped. The overvoltage relay closes breaker *B* when the preset voltage level is reached. The characteristics of undervoltage relays are indicated in Figure 9-15.

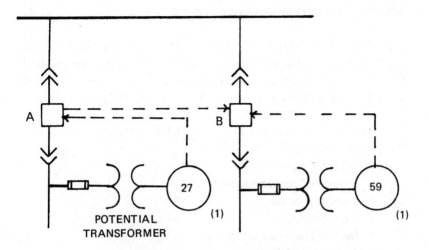

Figure 9-14. One-Line Diagram for
Undervoltage and Overvoltage Relay

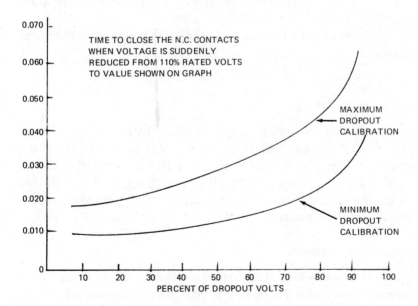

Figure 9-15. Characteristics of Undervoltage Relays

APPLYING SOLID-STATE PROTECTIVE RELAYS

Solid-state protective relays offer significantly improved characteristics over electromechanical relays and are available for the most important areas of the system. Solid-state techniques allow improvement in sensitivity and temperature stability, plus effective transient surge protection.

The main advantages in using solid-state relays are:
- *Flexible settings*
- *Dynamic performance*
- *Improved instantaneous*
- *Very low burden*
- *Easy testing*
- *Reduction in panel space*
- *Improved indication*
- *Better immunity to shock*
- *Good repeatability*

These relays are available in time overcurrent relays, instantaneous overcurrent relays, voltage relays, directional relays, reclosing relays, ground-fault relays, direct-current relays and timing relays.

Some of the advantages of solid-state relays are summarized below.

Easy Testing
And Flexible Settings

Installation testing is performed by depressing test buttons on the relay; thus test equipment is not required. These test buttons allow for initial settings, operational and wiring checks, and maintenance tests. A rough timing check can be performed with the second hand of an ordinary wrist watch. More precise tests can also be made.

Dynamic Performance

The solid-state relay can be thought of as a fine-tuned electromechanical protective device. The conventional electromechanical overcurrent relay is the induction disc type. The induction disc element which is either copper or aluminum rotates between the pole faces of an electromagnet. There are two methods commonly used to rotate the induction disc. The shaded pole method is illustrated in Figure 9-16. In this method a portion of the electromagnet pole face is short-circuited by a copper ring or coil to cause the flux in the shaded portion to lag the flux in the unshaded portion.

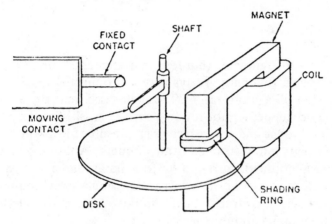

Figure 9-16. Shaded Pole Induction Disc Type Overcurrent Relay

The second method is referred to as the wattmetric type and uses one set of coils above and below the disc. In both methods the moving contact is carried on the rotating shaft. In the induction-disc type of overcurrent relay, the disc continues to rotate after the starting current has decreased to a low value. This overtravel means that in order to avoid nuisance tripping the time dial setting of the relay is set several positions higher than is desirable. Solid-state relays do not have rotating parts, thus the problem of overtravel does not occur. This means that the characteristics of the relay can be adjusted to what is required without allowances for dynamic effects such as overtravel.

The solid-state overcurrent relay consists of printed circuit boards that produce a d-c output voltage when the input a-c current exceeds a given value. Each overcurrent function usually consists of an input transformer, overcurrent module, and a resistor-zener-diode protective network. The overcurrent module consists of a setting circuit, phase-splitter circuit, sensing circuit, amplifier circuit, feedback circuit and an output circuit. The overcurrent module can either be a single-input module with one output or a dual-input module with a single output.

Low Burden

Low-volt ampere (burden) requirements for protective relays means that more relays may be connected in series, the lowest tap of a multi-ratio CT can be used, and the auxiliary CT can be stepped up for residual current sensitivity. This translates into dollar and space savings as a result of low-volt ampere requirements on instrument transformers. Another savings is that bush-mounted current transformers are not required with solid-state relays. Solid-state relays are compatible with bushing-mounted current transformers in low- and medium-voltage switchgear applications.

Reduction in Panel Space

A comparison of solid-state relay sizes with their electro-mechanical counterparts indicates a space savings of at least one-third. For complex relay schemes space savings can be 75 per cent or more.

Improved Indication

The solid-state protective relay has a target which operates independently of the trip-coil current and depends only on the proper functioning of the relay. This helps in troubleshooting in the event of a broken trip-coil connection. With the new design it is possible to have an indicator without a seal-in contact in parallel with the relay's measuring contacts. Often this results in simplification of complex schemes.

Better Immunity to Shock

Since the solid-state relay has no moving parts it is in many cases better suited for earthquake-prone locations. Solid-state relays have been tested and have withstood accelerations of up to 10g and higher.

Reliability

Solid-state devices are reliable, have good repeatability, and are economical for industrial applications. The major protective relay manufacturers now offer solid-state relays as part of their line. Over the last few years solid-state protective devices have proven to be a new tool for system protection.

Chapter 10

ENERGY ECONOMIC
ANALYSIS

To justify an energy investment cost, a knowledge of life-cycle costing is required.

The life-cycle cost analysis evaluates the total owning and operating cost. It takes into account the "time value" of money and can incorporate fuel cost escalation into the economic model. This approach is also used to evaluate competitive projects. In other words, the life-cycle cost analysis considers the cost over the life of the system rather than just the first cost.

THE TIME VALUE OF MONEY CONCEPT

To compare energy utilization alternatives, it is necessary to convert all cash flow for each measure to an equivalent base. The life-cycle cost analysis takes into account the "time value" of money, thus a dollar in hand today is more valuable than one received at some time in the future. This is why a time value must be placed on all cash flows into and out of the company.

DEVELOPING CASH FLOW MODELS

The cash flow model assumes that cash flows occur at discrete points in time as lump sums and that interest is computed and payable at discrete points in time.

To develop a cash flow model which illustrates the effect of "compounding" of interest payments, the cash flow model is developed as follows:

End of Year 1: $P + i(P) = (1 + i) P$

Year 2: $(1 + i)P + (1 + i)Pi = (1 + i)P [(1 + i)]$
$= (1 + i)^2 P$

Year 3: $(1 + i)^3 P$

Year n: $(1 + i)^n P$ or $S = (1 + i)^n P$

Where P = present sum
i = interest rate earned at the end of each interest period
n = number of interest periods
S = future value

$(1 + i)^n$ is referred to as the "Single Payment Compound Amount" factor and is tabluated for various values of i and n in Tables 1-8 in Appendix I.

The cash flow model can also be used to find the present value of a future sum S.

$$P = \left(\frac{1}{(1 + i)^n} \right) S$$

Cash flow models can be developed for a variety of other types of cash flow as illustrated in Figure 10-1.

To develop the Cash Flow Model for the "Uniform Series Compound Amount" factor, the following cash flow diagram is drawn.

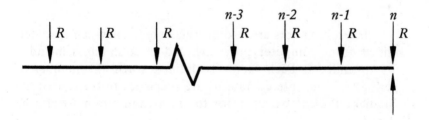

Where R is a Uniform Series of year-end payments and S is the future sum of R payments for n interest periods.

The R dollars deposited at the end of the nth period earn no interest and, therefore, contribute R dollars to the fund. The R dollars deposited at the end of the $(n-1)$ period earn interest for 1 year and will, therefore, contribute $R(1+i)$ dollars to the fund. The R dollars deposited at the end of the $(n-2)$ period earn interest for 2 years and will, therefore, contribute $R(1+i)^2$. These years of earned interest in the contributions will continue to increase in this manner, and the R deposited at the end of the first period will have earned interest for $(n-1)$ periods. The total in the fund S is, thus, equal to $R + R(1+i) + R(1+i)^2 + R(1+i)^3 + R(1+i)^4 + \ldots + R(1+i)^{n-2} + R(1+i)^{n-1}$. Factoring out R,

(1) $\quad S = R[1 + (1+i) + (1+i)^2 + \ldots + (1+i)^{n-2} + (1+i)^{n-1}]$

Multiplying both sides of this equation by $(1+i)$;

(2) $\quad (1+i)S = R[(1+i) + (1+i)^2 + (1+i)^3 + \ldots + (1+i)^{n-1} + (1+i)^n]$

Subtracting equation (1) from (2):

$$(1+i)S - S = R[(1+i) + (1+i)^2 + (1+i)^3 \ldots + (1+i)^{n-1} + (1+i)^n] - R[1 + (1+i) + (1+i)^2 + \ldots + (1+i)^{n-2} + (1+i)^{n-1}]$$

$$iS = R[(1+i)^n - 1]$$

$$S = R\left[\frac{(1+i)^n - 1}{i}\right]$$

Interest factors are seldom calculated. They can be determined from computer programs, and interest tables included in Appendix I. Each factor is defined when the number of periods (n) and interest rate (i) are specified. In the case of the Gradient Present Worth factor the escalation rate must also be stated.

The three most commonly used methods in life-cycle costing are the annual cost, present worth and rate-of-return analysis.

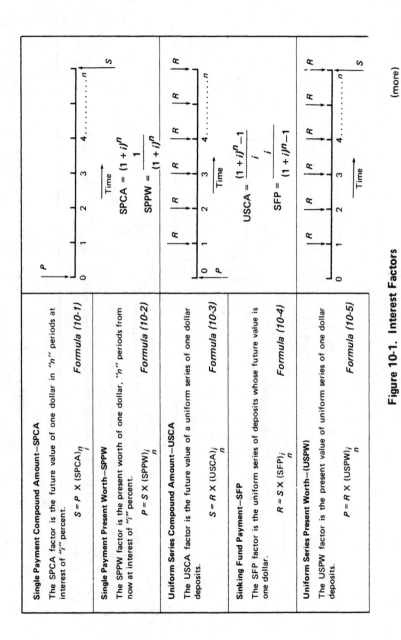

Single Payment Compound Amount—SPCA

The SPCA factor is the future value of one dollar in "n" periods at interest of "i" percent.

$$S = P \times (SPCA)^n_i \qquad \textit{Formula (10-1)}$$

$$SPCA = (1 + i)^n$$

Single Payment Present Worth—SPPW

The SPPW factor is the present worth of one dollar, "n" periods from now at interest of "i" percent.

$$P = S \times (SPPW)_i^n \qquad \textit{Formula (10-2)}$$

$$SPPW = \frac{1}{(1 + i)^n}$$

Uniform Series Compound Amount—USCA

The USCA factor is the future value of a uniform series of one dollar deposits.

$$S = R \times (USCA)_i^n \qquad \textit{Formula (10-3)}$$

$$USCA = \frac{(1 + i)^n - 1}{i}$$

Sinking Fund Payment—SFP

The SFP factor is the uniform series of deposits whose future value is one dollar.

$$R = S \times (SFP)_i^n \qquad \textit{Formula (10-4)}$$

$$SFP = \frac{i}{(1 + i)^n - 1}$$

Uniform Series Present Worth—(USPW)

The USPW factor is the present value of uniform series of one dollar deposits.

$$P = R \times (USPW)_i^n \qquad \textit{Formula (10-5)}$$

Figure 10-1. Interest Factors

(more)

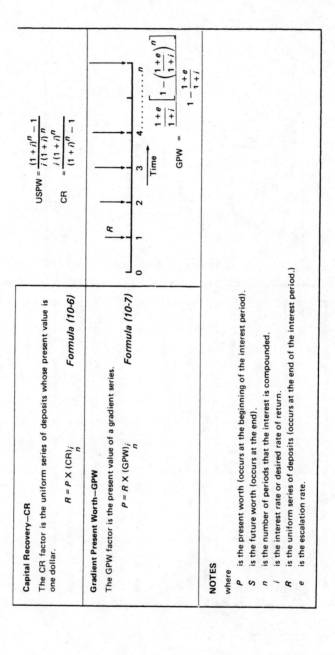

Capital Recovery—CR

The CR factor is the uniform series of deposits whose present value is one dollar.

$$R = P \times (CR)_{i}^{n} \qquad \text{Formula (10-6)}$$

Gradient Present Worth—GPW

The GPW factor is the present value of a gradient series.

$$P = R \times (GPW)_{i}^{n} \qquad \text{Formula (10-7)}$$

NOTES

where

P is the present worth (occurs at the beginning of the interest period).

S is the future worth (occurs at the end).

n is the number of periods that the interest is compounded.

i is the interest rate or desired rate of return.

R is the uniform series of deposits (occurs at the end of the interest period.)

e is the escalation rate.

$$USPW = \frac{(1+i)^n - 1}{i(1+i)^n}$$

$$CR = \frac{i(1+i)^n}{(1+i)^n - 1}$$

$$GPW = \frac{\frac{1+e}{1+i}\left[1 - \left(\frac{1+e}{1+i}\right)^n\right]}{1 - \frac{1+e}{1+i}}$$

Figure 10-1. Interest Factors (concluded)

In the present worth method a minimum rate of return (i) is stipulated. All future expenditures are converted to present values using the interest factors. The alternative with lowest effective first cost is the most desirable.

A similar procedure is implemented in the annual cost method. The difference is that the first cost is converted to an annual expenditure. The alternative with lowest effective annual cost is the most desirable.

In the rate-of-return method, a trial-and-error procedure is usually required. Interpolation from the interest tables can determine what rate of return (i) will give an interest factor which will make the overall cash flow balance. The rate-of-return analysis gives a good indication of the overall ranking of independent alternates.

The effect of escalation in fuel costs can influence greatly the final decision. When an annual cost grows at a steady rate it may be treated as a gradient and the Gradient Present Worth factor can be used.

Special appreciation is given to Rudolph R. Yaneck and Dr. Robert Brown for use of their specially designed interest and escalation tables used in this text.

When life-cycle costing is used to compare several alternatives the differences between costs are important. For example, if one alternate forces additional maintenance or an operating expense to occur, then these factors as well as energy costs need to be included. Remember, what was previously spent for the item to be replaced is irrelevant. The only factor to be considered is whether the new cost can be justified based on projected savings over its useful life.

PAYBACK ANALYSIS

The simple payback analysis is sometimes used instead of the methods previously outlined. The simple payback is defined as initial investment divided by annual savings after taxes. The simple payback method does not take into account the effect of interest or escalation rate.

Since the payback period is relatively simple to calculate and due to the fact managers wish to recover their investment as rapidly as possible the payback method is frequently used.

It should be used in conjunction with other decision-making tools. When used by itself as the principal criterion it may result in choosing less profitable investments which yield high initial returns for short periods as compared with more profitable investments which provide profits over longer periods of time.

SIM 10-1

An electrical energy audit indicates electrical motor consumption is 4×10^6 KWH per year. By upgrading the motor spares with high efficiency motors a 10% savings can be realized. The additional cost for these motors is estimated at $80,000. Assuming an 8¢ per KWH energy charge and 20-year life, is the expenditure justified based on a minimum rate of return of 20% before taxes? Solve the problem using the present worth, annual cost, and rate-of-return methods.

Analysis

Present Worth Method

	Alternate 1 *Present Method*	*Alternate 2* *Use High Efficiency* *Motor Spares*
(1) First Cost *(P)*	—	$80,000
Annual Cost *(R)*	$4 \times 10^{12} \times .08$	$.9 \times \$320,000$
	= $320,000	= $288,000
USPW (Table 4, App. I)	4.87	4.87
(2) *R* X USPW =	$1,558.400	$1,402,560
Present Worth	$1,558,400	$1,482,560
(1) + (2)		Choose Alternate with Lowest First Cost

Annual Cost Method

	Alternate 1	Alternate 2
(1) First Cost *(P)*	—	$80,000
Annual Cost *(R)*	$320,000	$288,000
CR (Table 16-4)	.2	.2
(2) *P* X CR	—	$16,000
Annual Cost	$320,000	$304,000
(1) + (2)		∠ Choose Alternate with Lowest First Cost

Rate of Return Method

$$P = R\,(USPW) = (\$320,000 - \$288,000) \times USPW$$

$$USPW = \frac{80,000}{32,000} = 2.5$$

What value of *i* will make USPW = 2.5? *i* = 40% (Table 7, App. I).

SIM 10-2

Show the effect of 10% escalation on the rate-of-return analysis given the

Energy equipment investment = $20,000
After tax savings   = $ 2,600
Equipment life *(n)*   = 15 years

Analysis

Without escalation

$$CR = \frac{R}{P} = \frac{2,600}{20,000} = .13$$

From Table 1, App. I, the rate of return is 10%.
With 10% escalation assumed:

$$GPW = \frac{P}{G} = \frac{20,000}{2,600} = 7.69$$

From Table 11, App. I, the rate of return is 21%.

Thus we see that taking into account a modest escalation rate can dramatically affect the justification of the project.

DEPRECIATION, TAXES, AND THE TAX CREDIT

Depreciation

Depreciation affects the "accounting procedure" for determining profits and losses and the income tax of a company. In other words, for tax purposes the expenditure for an asset such as a pump or motor can not be fully expensed in its first year. The original investment must be charged off for tax purposes over the useful life of the asset. A company usually wishes to expense an item as quickly as possible.

The Internal Revenue Service allows several methods for determining the annual depreciation rate.

Straight-Line Depreciation: The simplest method is referred to as a straight-line depreciation and is defined as:

$$D = \frac{P - L}{n} \qquad \text{\textit{Formula (10-8)}}$$

Where:

D is the annual depreciation rate

L is the value of equipment at the end of its useful life, commonly referred to as salvage value

n is the life of the equipment which is determined by Internal Revenue Service Guidelines

P is the initial expenditure.

Sum-of-Years Digits: Another method is referred to as the sum-of-years digits. In this method the depreciation rate is determined by finding the sum of digits using the following formula:

$$N = n\,\frac{(n + 1)}{2} \qquad \text{\textit{Formula (10-9)}}$$

Where n is the life of equipment.

Each year's depreciation rate is determined as follows:

First year
$$D = \frac{n}{N}(P - L)$$
Formula (10-10)

Second year
$$D = \frac{n-1}{N}(P - L)$$
Formula (10-11)

n year
$$D = \frac{1}{N}(P - L)$$
Formula (10-12)

Declining-Balance Depreciation: The declining-balance method allows for larger depreciation charges in the early years which is sometimes referred to as fast write-off.

The rate is calculated by taking a constant percentage of the declining undepreciated balance. The most common method used to calculate the declining balance is to predetermine the depreciation rate. Under certain circumstances a rate equal to 200% of the straight-line depreciation rate may be used. Under other circumstances the rate is limited to 1½ or ¼ times as great as straight-line depreciation. In this method the salvage value or undepreciated book value is established once the depreciation rate is preestablished.

To calculate the undepreciated book value Formula 10-13 is used:

$$D = 1 - \left(\frac{L}{P}\right)^{1/N}$$
Formula (10-13)

Where
    D is the annual depreciation rate
    L is the salvage value
    P is the first cost.

SIM 10-3

Calculate the depreciation rate using the straight-line, sum-of-years digit, and declining-balance methods.
    Salvage value is 0
    $n = 5$ years

$P = 150,000$
For declining balance use a 200% rate.

Straight-Line Method

$$D = \frac{P - L}{n} = \frac{150,000}{5} = \$30,000 \text{ per year}$$

Sum-of-Years Digits

$$N = \frac{n(n + 1)}{2} = \frac{5(6)}{2} = 15$$

$$D_1 = \frac{n}{N}(P) = \frac{5}{15}(150,000) = 50,000$$

N	P
1 =	$50,000
2 =	40,000
3 =	30,000
4 =	20,000
5 =	10,000

Declining-Balance Method

$D = 2 \times 20\% = 40\%$ (Straight-Line Depreciation Rate = 20%)

Year	Undepreciated Balance at Beginning of Year	Depreciation Charge
1	150,000	60,000
2	90,000	36,000
3	54,000	21,600
4	32,400	12,960
5	19,440	7,776
	TOTAL	138,336

Undepreciated Book Value (150,000 − 138,336) = $11,664

Economic Recovery Tax Act—1981

The Economic Recovery Tax Act allows for an accelerated depreciation over a shorter life. Thus energy investments will become more attractive.

Tax Considerations

Tax-deductible expenses such as maintenance, energy, operating costs, insurance and property taxes reduce the income subject to taxes.

For the after-tax life-cycle cost analysis and payback analysis the actual incurred annual savings is given as follows:

$$AS = (1-I)\,E + ID \qquad\qquad Formula\ (10\text{-}14)$$

Where:

AS = yearly annual after-tax savings (excluding effect of tax credit)

E = yearly annual energy savings (difference between original expenses and expenses after modification)

D = annual depreciation rate

I = income tax bracket

Formula 10-14 takes into account that the yearly annual energy savings is partially offset by additional taxes which must be paid due to reduced operating expenses. On the other hand, the depreciation allowance reduces taxes directly.

Tax Credit

A tax credit encourages capital investment. Essentially the tax credit lowers the income tax paid by the tax credit to an upper limit.

In addition to the investment tax credit, the Business Energy Tax Credit as a result of the National Energy Plan, can also be taken. The Business Energy Tax Credit applies to industrial investment in alternative energy property such as boilers for coal, heat conservation, and recycling equipment. The tax credit substantially increases the investment merit of the investment since it lowers the *bottom* line on the tax form. Since tax laws are in constant flux, check to determine the extent tax credits are in effect at the time of the analysis.

After-Tax Analysis

To compute a rate of return which accounts for taxes, depreciation, escalation, and tax credits, a cash-flow analysis is usually required. This method analyzes all transactions including first and operating costs. To determine the after-tax rate of return a trial and error or computer analysis is required.

The Present Worth factors tables in Appendix I can be used for this analysis. All money is converted to the present assuming an interest rate. The summation of all present dollars should equal zero when the correct interest rate is selected, as illustrated in Figure 10-2.

This analysis can be made assuming a fuel escalation rate by using the Gradient Present Worth interest of the Present Worth Factor.

SIM 10-4

Comment on the after-tax rate of return for the installation of a heat-recovery system with and without tax credit given the following:

- First Cost $100,000
- Year Savings   40,000
- Straight-line depreciation life and equipment life of 5 years
- Income tax bracket 46%

Analysis

$$D = 100,000/5 = 20,000$$

$$AS = (1-I)E + ID = .54(40,000) + .46(20,000)$$
$$= 21,600 + 9,200 = 30,800$$

Year	1 Investment	2 Tax Credit	3 After Tax Savings (AS)	4 Single Payment Present Worth Factor	(2 + 3) X 4 Present Worth
0	−P				−P
1		+TC	AS_1	$SPPW_1$	+P_1
2			AS_2	$SPPW_2$	P_2
3			AS_3	$SPPW_3$	P_3
4			AS_4	$SPPW_4$	P_4
Total					ΣP

$$AS = (1-I)E + ID$$
Trial & Error Solution:
Correct i when $\Sigma P = 0$

Figure 10-2. Cash Flow Rate of Return Analysis

Without Tax Credit
First Trial i = 20%

Investment	After Tax Savings	SPPW 20%	PW
0−100,000			−100,000
1	30,800	.833	25,656
2	30,800	.694	21,375
3	30,800	.578	17,802
4	30,800	.482	14,845
5	30,800	.401	12,350
		$\Sigma-$	7,972

Since summation is negative a higher present worth factor is required. Next try is 15%.

Investment	After Tax Savings	SPPW 15%	PW
0–100,000			−100,000
1	30,800	.869	+ 26,765
2	30,800	.756	+ 23,284
3	30,800	.657	+ 20,235
4	30,800	.571	+ 17,586
5	30,800	.497	+ 15,307
			+ 3,177

Since rate of return is bracketed, linear interpolation will be used.

$$\frac{3177 + 7971}{-5} = \frac{3177-0}{15-i\%}$$

$$i = \frac{3177}{2229.6} + 15 = 16.4\%$$

With Tax Credit

Tax Credit = 10% (Investment) + 10% (Energy) = 20%

Investment	Tax Credit	After Tax Savings	SPPW 20%	PW
0–100,000				−100,000
1	20,000	30,800	.833	45,656
2		30,800	.694	21,375
3		30,800	.578	17,802
4		30,800	.482	14,845
5		30,800	.401	12,350
				+ 12,028

Next try 25%.

$$PW$$

$$
\begin{array}{r}
-100,000 \\
44,640 \\
19,712 \\
15,769 \\
10,093 \\
\hline
-9,786
\end{array}
$$

$$\frac{12,028 + 9,786}{-5} = \frac{12,028}{20-i}$$

$$i = 22.75\%$$

THE IMPACT OF FUEL INFLATION
ON THE LIFE CYCLE ANALYSIS

The rate of return on investment becomes more attractive when life cycle costs are taken into account. Tables 9 through 12 can be used to show the impact of fuel inflation on the decision-making process.

SIM 10-5

Develop a set of curves that indicate the capital that can be invested to give a rate of return of 15% after taxes for each $1000 saved for the following conditions.

1. The effect of escalation is not considered.
2. A 5% fuel escalation is considered.
3. A 10% fuel escalation is considered.
4. A 14% fuel escalation is considered.
5. A 20% fuel escalation is considered.

Calculate for 5-, 10-, 15-, 20-year life.

Assume straight-line depreciation over useful life, 48% income tax bracket, and no tax credit.

Answer

$$AS = (1-I)E + ID$$

$$I = 0.48, E = \$1000$$

$$AS = 520 + \frac{0.48P}{N}$$

Thus, the after-tax savings (AS) are comprised of two components. The first component is a uniform series of $520 escalating at e percent/year. The second component is a uniform series of $0.48P/N$.

Each component is treated individually and converted to present-day values using the GPW factor and the USPW factor, respectively. The sum of these two present worth factors must equal P. In the case of no escalation, the formula is:

$$P = 520 \text{ USPW} + \frac{0.48P}{n} \text{ USPW.}$$

In the case of escalation:

$$P = 520 \text{ GPW} + \frac{0.48P}{n} \text{ USPW.}$$

Since there is only one unknown, the formulas can be readily solved. The results are indicated below.

	n = 5 $P	n = 10 $P	n = 15 $P	n = 20 $P
$e = 0$	2,570.79	3,434	3,753.62	3,829.03
$e = 10\%$	3,362.35	5,401.84	6,872.98	7,922.35
$e = 14\%$	3,735.87	6,523.26	8,986.38	11,176.47
$e = 20\%$	4,364.30	8,710.68	\$13,728.27	\$19,698.82

Figure 10-3 illustrates the effects of escalation. This figure can be used as a quick way to determine after-tax economics of energy utilization expenditures.

SIM 10-6

It is desired to have an after-tax savings of 15%. Comment on the investment that can be justified if it is assumed that the fuel rate escalation should not be considered and the annual energy savings is $2000 with an equipment economic life of 12 years.

Comment on the above, assuming a 14% fuel escalation.

Answer

From Figure 10-3, for each $1000 energy savings, an investment of $3600 is justified or $7200 for a $2000 savings when no fuel increase is accounted for.

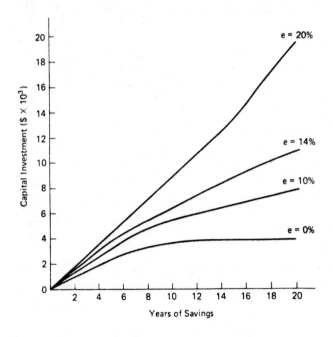

Figure 10-3. Effects of Escalation

***** EER ENERGY PRICE FORECASTS *****

JANUARY 1983

REGION   UNITED STATES

UPDATES EVERY SIX MONTHS

	INDUSTRIAL ELECTRICITY			INDUSTRIAL DISTILLATE FUEL			INDUSTRIAL NATURAL GAS		
	CURRENT $ PER MM BTU	CURRENT $ PER KWH	ESCALATION RATE*	CURRENT $ PER MM BTU	CURRENT $ PER GAL	ESCALATION RATE*	CURRENT $ PER MM BTU	CURRENT $ PER MCF	ESCALATION RATE*
1980	10.30	0.04		6.67	0.93		2.76	2.92	
1981	11.81	0.04		7.80	1.08		3.24	3.31	
1982	13.07	0.04		7.59	1.05		3.78	3.86	
1983	14.01	0.05	BASE	8.04	1.12	BASE	4.62	4.72	BASE
1984	15.89	0.05	13.4	8.57	1.19	6.6	5.22	5.33	13.0
1985	17.93	0.06	13.1	9.12	1.26	6.5	6.40	6.53	17.7
1986	19.37	0.07	11.4	10.23	1.42	8.4	7.12	7.27	15.5
1987	20.92	0.07	10.5	11.48	1.59	9.3	7.93	8.09	14.5
1988	22.60	0.08	10.0	12.88	1.79	9.9	8.82	9.01	13.4
1989	24.41	0.08	9.7	14.45	2.00	10.3	9.82	10.03	13.1
1990	26.37	0.09	9.5	16.21	2.25	10.5	10.93	11.16	13.1
1991	28.05	0.10	9.1	17.89	2.48	10.5	12.08	12.34	12.8
1992	29.84	0.10	8.8	19.73	2.74	10.0	13.36	13.64	12.5
1993	31.74	0.11	8.5	21.78	3.02	10.0	14.77	15.07	12.2
1994	33.76	0.11	8.3	24.07	3.33	10.0	16.33	16.63	12.0
1995	35.91	0.12	8.2	26.51	3.68	10.5	18.05	18.43	12.0
1996	37.87	0.13	7.9	28.27	3.92	10.2	19.25	19.66	11.6
1997	39.93	0.14	7.8	30.15	4.18	9.7	20.60	20.96	11.2
1998	42.11	0.14	7.6	32.16	4.46	9.5	21.90	22.36	10.7
1999	44.40	0.15	7.5	34.30	4.76	9.3	23.36	23.85	10.4
2000	46.82	0.16	7.4	36.58	5.07	9.3	24.91	25.43	10.4

* AVERAGE ANNUAL ESCALATION RATE OF PRICES, PERCENT FROM BASE YEAR TO CURRENT YEAR

COPYRIGHT 1983

Source: *EER Energy Price Forecasts, PO Box 14227, Atlanta, GA 30324*

Figure 10-4. Typical Fuel Price Projections

With a 14% fuel escalation rate, an investment of $6000 is justified for each $1000 energy savings, thus $12,000 can be justified for $2000 savings. Thus, a 66% higher expenditure is economically justifiable and will yield the same after-tax rate of return of 15% when a fuel escalation of 14% is considered.

In order to estimate future fuel costs a computer simulation, forecasting service or historical data may be used. One such forecast is illustrated in Figure 10-4. It should be noted that even though larger price projections are made in the short term the average long-term projections are in the order of 7 to 10% depending on the fuel type. The average escalation number from this figure is then used as illustrated in the previous examples.

Chapter 11

ENERGY MANAGEMENT
SYSTEMS

The availability of computers at moderate costs and concern for reducing energy consumption has resulted in the application of computer-based controllers to more than just industrial process applications. These controllers, commonly called Energy Management Systems (EMS), can be used to control virtually all non-process energy using pieces of equipment in buildings and industrial plants. Equipment controlled can include fans, pumps, boilers, chillers and lights. This chapter will investigate the various types of Energy Management Systems which are available and illustrate some of the methods used to reduce energy consumption.

THE TIMECLOCK

One of the simplest and most effective methods of conserving energy in a building is to operate equipment only when it is needed. If due to time, occupancy, temperature or other means it can be determined that a piece of equipment does not need to operate, energy savings can be achieved without affecting occupant comfort by turning the equipment off.

One of the simplest devices to schedule equipment operation is the mechanical timeclock. The timeclock consists of a rotating disk which is divided into segments corresponding to the hour of the day and the day of the week. This disk makes one complete revolution in, depending on the type, a 24-hour or a 7-day period. See Figure 11-1.

On and off "lugs" are attached to the disk at appropriate positions corresponding to the schedule for the piece of equipment. As the disk rotates, the lugs cause a switch contact to open and close thereby controlling equipment operation.

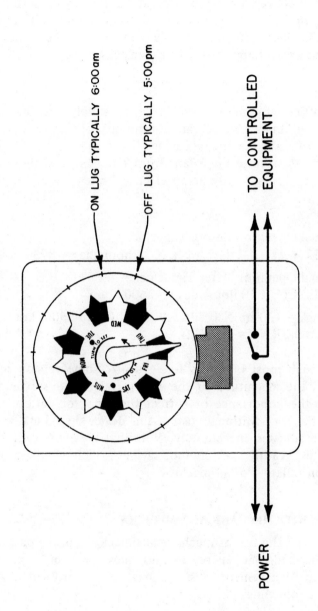

Figure 11-1. Mechanical Timeclock

A common application of timeclocks is scheduling office building HVAC equipment to operate during business hours Monday thru Friday and to be off all other times. As is shown in the following problem, significant savings can be achieved through the correct application of timeclocks.

SIM 11-1

An office building utilizes two 50 hp supply fans and two 15 hp return fans which operate continuously to condition the building. What are the annual savings that result from installing a timeclock to operate these fans from 7:00 a.m. to 5:00 p.m., Monday thru Friday? Assume an electrical rate of $0.08/KWH.

Answer

Annual Operation Before Timeclock =
52 weeks X 7 days/week X 24 hours/day = 8736 hours

Annual Operation After Timeclock =
52 X (5 X 10 hours/day) = 2600 hours

Savings = 130 hp X 0.746 KW/hp X (8736-2600) hours X $0.08/KWH = $47,600

Although most buildings today utilize some version of a timeclock, the magnitude of the savings value in this example illustrates the importance of correct timeclock operation and the potential for additional costs if this device should malfunction or be adjusted inaccurately. Note that the above example also ignores heating and cooling savings which would result from the installation of a timeclock.

PROBLEMS WITH MECHANICAL TIMECLOCKS

Although the use of mechanical timeclocks in the past has resulted in significant energy savings, they are being replaced by Energy Management Systems because of problems that include the following.

- The on/off lugs sometimes loosen or fall off.

- Holidays, when the building is unoccupied, cannot easily be taken into account.

- Power failures require the timeclock to be reset or it is not synchronized with the buiding schedule.

- Inaccuracies in the mechanical movement of the time-clock prevent scheduling any closer than plus or minus 15 minutes of the desired times.

- There are a limited number of on and off cycles possible each day.

- It is a time-consuming process to change schedules on multiple timeclocks.

Energy Managment Systems, or sometimes called electronic timeclocks, are designed to overcome these problems plus provide increased control of building operations.

ENERGY MANAGEMENT SYSTEMS

Recent advances in digital technology, dramatic decreases in the cost of this technology and increased energy awareness has resulted in the increased application of computer-based controllers (i.e., Energy Management Systems) in commercial buildings and industrial plants. These devices can control any-where from one to a virtually unlimited number of items of equipment.

By concentrating the control of many items of equipment at a single point, the EMS allows the building operator to tailor building operation to precisely satisfy occupant needs. This ability to maximize energy conservation, while preserving occupant comfort, is the ultimate goal of an energy engineer.

Microprocessor Based

Energy Management Systems can be placed in one of two broad, and sometime overlapping, categories referred to as microprocessor-based and mini-computer based.

Microprocessor-based systems can control from 1 to 50 (approximately) separate points or loads. (By a control point is meant an item of equipment which is operated according to a unique schedule). Programming is accomplished by a keyboard on the front panel and a LED or LCD display is used to monitor/review operation of the unit. A battery maintains the programming in the event of power failure. See Figure 11-2.

Capabilities of this type of EMS are generally pre-programmed so that operation is relatively straightforward. Programming simply involves entering the appropriate parameters (eg. the point number and the on and off times) for the desired function. Microprocessor-based EMS can have any or all of the following capabilities:

- Scheduling
- Duty Cycling
- Demand Limiting
- Optimal Start
- Monitoring
- Direct Digital Control

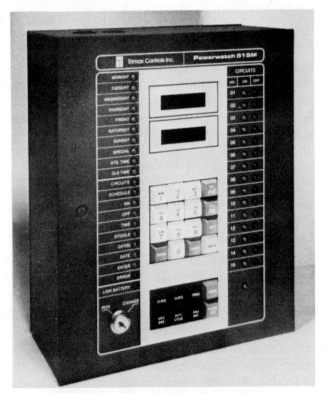

Figure 11-2. Microprocessor-Based EMS*

*Courtesy of Trimax Division of Margaux Controls.

SCHEDULING

Scheduling with an EMS is very much the same as it is with a timeclock. Equipment is started and stopped based on the time of day and the day of week. Unlike a timeclock, however, multiple start/stops can be accomplished very easily and accurately (eg. In a classroom, lights can be turned off during morning and afternoon break periods and during lunch.). It should be noted that this single function, if accurately programmed and depending on the type of facility served, can account for the largest energy savings attributable to an EMS.

Additionally, holiday dates can be entered into the EMS a year in advance. When the holiday occurs, regular programming is overridden and equipment can be kept off.

DUTY CYCLING

Most HVAC fan systems are designed for peak load conditions, and consequently these fans are usually moving much more air than is needed. Therefore, they can sometimes be shut down for short periods each hour, typically 15 minutes, without affecting occupant comfort. Turning equipment off for pre-determined periods of time during occupied hours is referred to as **duty cycling**, and can be accomplished very easily with an EMS. Duty cycling saves fan and pump energy, but does not reduce the energy required for space heating or cooling since the thermal demand must still be met.

The more sophisticated EMS's monitor the temperature of the conditioned area and use this information to automatically modify the duty cycle length when temperatures begin to drift. If for example, the desired temperature in an area is 70° and at this temperature equipment is cycled 50 minutes on and 10 minutes off, a possible temperature compensated EMS may respond as shown in Figure 11-3. As the space temperature increases above (or below if so programmed) the setpoint, the equipment off time is reduced until, at 80°, in this example, the equipment operates continuously.

Duty cycling is best applied in large open space offices which are served by a number of fans. Each fan could be programmed so that the off times do not coincide, thereby assuring adequate air flow to the offices at all times.

Duty cycling of fans which provide the only air flow to an area should be approached carefully to insure that ventilation requirements are maintained and that varying equipment noise does not annoy the occupants. Additionally, duty cycling of equipment imposes extra stress on motors and associated equipment. Care should be taken, particularly with motors over 20 hp, to prevent starting and stopping of equipment in excess of what is recommended by the manufacturer.

DEMAND CHARGES

Electrical utilities charge commercial customers based not only on the amount of energy used (KWH) but also on the peak demand (KW) for each month. Peak demand is very important to the utility so that they may properly size the required electrical service and insure that sufficient peak generating capacity is available to that given facility.

In order to determine the peak demand during the billing period, the utility establishes short periods of time called the demand interval (typically 15, 30, or 60 minutes). The billing demand is defined as the highest average demand recorded during any one demand interval within the billing period. (See Figure 11-4.) Many utilities now utilize "ratchet" rate charges. A "ratchet" rate means that the billed demand for the month is based on the highest demand in the previous 12 months, or an average of the current month's peak demand and the previous highest demand in the past year.

Depending on the facility, the demand charge can be a significant portion, as much as 20%, of the utility bill. The user will get the most electrical energy per dollar if the load is kept constant, thereby minimizing the demand charge. The objective of demand control is to even out the peaks and valleys of consumption by deferring or rescheduling the use of energy during peak demand periods.

A measure of the electrical efficiency of a facility can be found by calculating the load factor. The load factor is defined

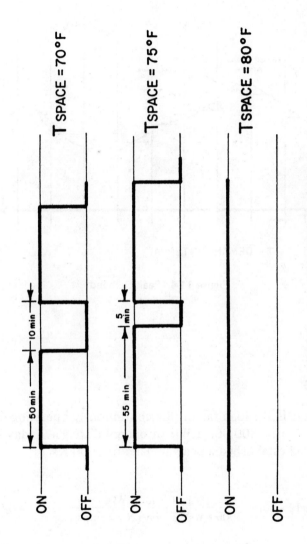

Figure 11-3. Temperature Compensated Cuty Cycling

as the ratio of energy usage (KWH) per month to the peak demand (KW) X the facility operating hours.

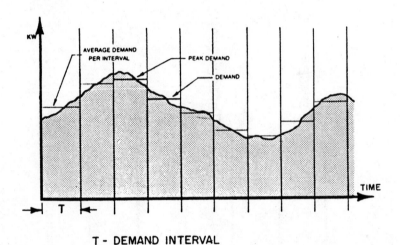

T - DEMAND INTERVAL

Figure 11-4. Peak Demand

SIM 11-2

What is the load factor of a continuously operating facility that consumed 800,000 KWH of energy during a 30-day billing period and established a peak demand of 2000 KW?

Answer

$$\text{Load Factor} = \frac{800{,}000 \text{ KWH}}{2000 \text{ KW X 30 days X 24 hours/day}} = 0.55$$

The ideal load factor is 1.0, at which demand is constant; therefore, the difference between the calculated load factor and 1.0 gives an indication of the potential for reducing peak demand (and demand charges) at a facility.

DEMAND LIMITING

Energy management systems with **demand limiting** capabilities utilize either pulses from the utility meter or current transformers to predict the facility demand during any demand interval. If the facility demand is predicted to exceed the user-entered setpoint, equipment is "shed" to control demand. Figure 11-5 illustrates a typical demand chart before and after the actions of a demand limiter.

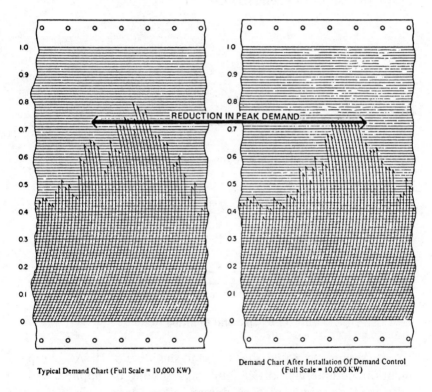

Typical Demand Chart (Full Scale = 10,000 KW)

Demand Chart After Installation Of Demand Control
(Full Scale = 10,000 KW)

Figure 11-5. Demand Limiting Comparison

Electrical load in a facility consists of two major categories: essential loads which include most lighting, elevators, escalators, and most production machinery; and non-essential ("sheddable") loads such as electric heaters, air conditioners, exhaust fans, pumps, snow melters, compressors and water heaters. Sheddable loads will not, when turned off for short periods of time to control demand, affect productivity or comfort.

To prevent excessive cycling of equipment, most energy management systems have a deadband that demand must drop below before equipment operation is restored (See Figure 11-6). Additionally, minimum on and maximum off times and shed priorities can be entered for each load to protect equipment and insure that comfort is maintained.

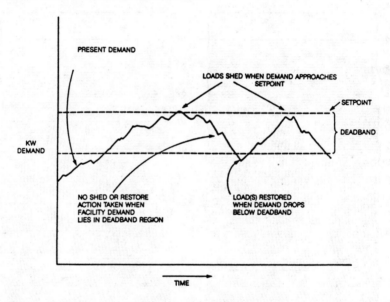

Figure 11-6. Demand Limiting Actions

It should be noted that demand shedding of HVAC equipment in commercial office buildings should be applied with caution. Since times of peak demand often occur during times of peak air conditioning loads, excessive demand limiting can result in occupant discomfort.

TIME OF DAY BILLING

Many utilities are beginning to charge their larger commercial users based on the **time of day** that consumption occurs.

Energy and demand during peak usage periods (i.e., summer weekday afternoons and winter weekday evenings) are billed at much higher rates than consumption during other times. This is necessary because utilities must augment the power production of their large power plants during periods of peak demand with small generators which are expensive to operate. Some of the more sophisticated energy management systems can now account for these peak billing periods with different demand setpoints based on the time of day and day of week.

OPTIMAL START

External building temperatures have a major influence on the amount of time it takes to bring the building temperature up to occupied levels in the morning. Buildings with mechanical time clocks usually start HVAC equipment operation at an early enough time in the morning (as much as 3 hours before occupancy time) to bring the building up to temperature on the coldest day of the year. During other times of the year when temperatures are not as extreme, building temperatures can be up to occupied levels several hours before it is necessary, and consequently unnecessary energy is used. See Figure 11-7.

Energy management systems with **optimal start** capabilities, however, utilize indoor and outdoor temperature information, along with learned building characteristics, to vary start time of HVAC equipment so that building temperatures reach desired values just as occupancy occurs. Consequently, if a building is scheduled to be occupied at 8:00 a.m., on the coldest day of the year, the HVAC equipment may start at 5:00 a.m. On milder days, however, equipment may not be started until 7:00 a.m. or even later, thereby saving significant amounts of energy.

Most energy management systems have a "self-tuning" capability to allow them to learn the building characteristics. If the building is heated too quickly or too slowly on one day, the start time is adjusted the next day to compensate.

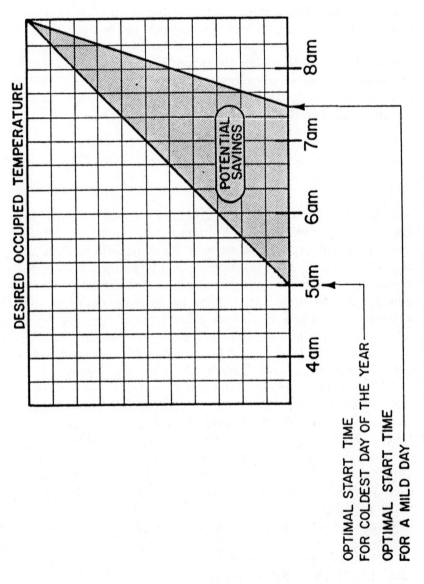

Figure 11-7. Typical Variation in Building Warm-up Times

MONITORING

Microprocessor-based EMS can usually accomplish a limited amount of **monitoring** of building conditions including the following:

- Outside air temperature
- Several indoor temperature sensors
- Facility electrical energy consumption and demand
- Several status input points

The EMS can store this information to provide a history of the facility. Careful study of these trends can reveal information about facility operation that can lead to energy conservation strategies that might otherwise not be apparent.

DIRECT DIGITAL CONTROL

The most sophisticated of the micro-processor-based EMS's provide a function referred to as **direct digital control** (DDC). This capability allows the EMS to provide not only sophisticated energy management but also basic temperature control of the building's HVAC systems.

Direct digital control has taken over the majority of all process control applications, and is now beginning to become an important part of the HVAC industry. Traditionally, pneumatic controls were used in most commercial facilities for environmental control.

The control function in a traditional facility is performed by a pneumatic controller which receives its input from pneumatic sensors (i.e., temperature, humidity), and sends control signals to pnaumatic actuators (valves, dampers, etc.). Pneumatic controllers typically perform a single, fixed function which cannot be altered unless the controller itself is changed or other hardware is added. See Figure 11-8 for a typical pneumatic control configuration.

With direct digital control, the microprocessor functions as the primary controller. Electronic sensors are used to measure variables such as temperature, humidity, and pressure. This information is used, along with the appropriate application program, by the microprocessor to determine the correct control signal, which is then sent directly to the controlled device (valve or damper actuator). See Figure 11-8 for a typical DDC configuration.

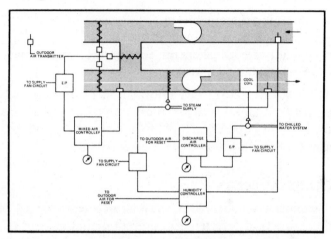

CONVENTIONAL PNEUMATIC CONTROL SYSTEM

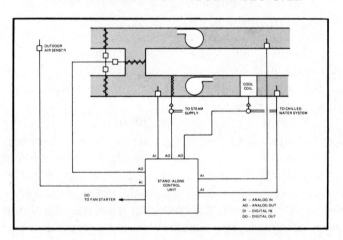

DIRECT DIGITAL CONTROL SYSTEM

Figure 11-8. Comparison of Pneumatic and DDC Controls

Direct Digital Control (DDC) has the following advantages over pneumatic controls:

- Reduces overshoot and offset errors, thereby saving energy.
- Flexibility to easily and inexpensively accomplish changes of control strategies.
- Calibration is maintained more accurately, thereby saving energy and providing better performance.

To program the DDC functions, a user programming language is utilized. This programming language uses simple commands in English to establish parameters and control strategies.

MINI-COMPUTER BASED

Mini-computer based EMS can provide all of the functions of the micro-processor based EMS, as well as the following:

- Extensive graphics
- Special reports and studies
- Fire and security monitoring and detection
- Custom programs

These devices can control and monitor from fifty (50) to an unlimited number of points and form the heart of a building's (or complex's) operations.

Figure 11-9 shows a typical configuration for this type of system.

The "central processing unit" (CPU) is the heart of the EMS. It is a minicomputer with memory for the operating system and applications software. The CPU performs arithmetic and logical decisions necessary to perform central monitoring and control.

Data and programs are stored or retrieved from the memory or mass storage devices (generally a disk storage system). The CPU has programmed I/O ports for specific equipment,

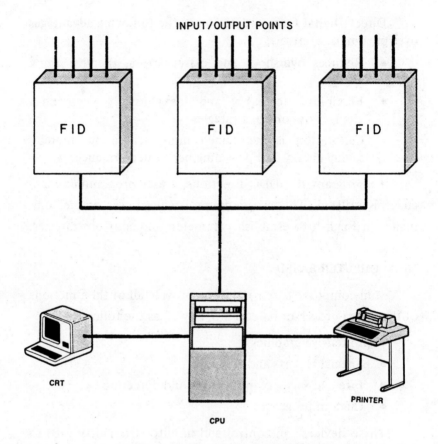

Figure 11-9. Mini-Computer-Based EMS

such as printers and cathode ray tube (CRT) consoles. During normal operation, it coordinates operation of all other EMS components.

A cathode ray tube console (CRT) either color and/or black and white, with a keyboard is used for operator interaction with the EMS. It accepts operator commands, displays data and graphically displays systems controlled or monitored by the EMS. A "printer," or printers, provides a permanent copy of system operations and historical data.

A "field interface device" (FID) provides an interface to the points which are monitored and controlled, performs engineering conversions to or from a digital format, performs calculations and logical operations, accepts and processes CPU commands and is capable, in some versions, of stand-alone operations in the event of CPU or communications link failure.

The FID is essentially a micro-processor based EMS as described in the previous section. It may or may not have a keyboard/display unit on the front panel.

The FID's are generally located in the vicinity of the points to be monitored and/or controlled and are linked together and to the CPU by a single twisted pair of wires which carries multiplexed (i.e., data from a number of sources combined on a single channel) data from the FID to the CPU and back. In some versions, the FID's can communicate directly with each other.

Early versions of mini-computer based EMS used the CPU to perform all of the processing with the FID used merely for input and output. A major disadvantage of this type of "centralized" system is that the loss of the CPU disables the entire control system. The development of "intelligent" FID's in a configuration known as "distributed processing" helped to solve this problem. This system, which is becoming prevalent today, utilizes microprocessor-based FID's to function as remote CPU's. Each panel has its own battery pack to insure continued operation should the main CPU fail.

Each intelligent panel sends signals back to the main CPU only upon a change of status rather than continuously transmitting the same value as previous "centralized" systems have done. This streamlining of data flow to the main CPU frees it to perform other functions such as trend reporting. The CPU's primary function becomes one of directing communications between various FID panels, generating reports and graphics and providing operator interface for programming and monitoring.

FEATURES

The primary difference in operating functions of the mini-computer based EMS is its increased capability to monitor building operations. For this reason, these systems are sometimes referred to as **Energy Monitoring and Control Systems** (EMCS). Analog inputs such as temperature and humidity can be monitored, as well as digital inputs such as pump or valve status.

The mini-computer based EMS is also designed to make operator interaction very easy. Its operation can be described as "user friendly" in that the operator, working through the keyboard, enters information in English in a question and response format. In addition, custom programming languages are available so that powerful programs can be created specifically for the building through the use of simplified English commands.

The graphics display CRT can be used to create HVAC schematics, building layouts, bar charts, etc. to better understand building systems operation. These graphics can be "dynamic" so that values and statuses are continuously updated.

Many mini-computer based EMS can also easily incorporate fire and security monitoring functions. Such a configuration is sometimes referred to as a **Building Automation System** (BAS). By combining these functions with energy management, savings in initial equipment costs can be achieved. Reduced operating costs can be achieved as well by having a single operator for these systems.

The color graphics display can be particularly effective in pinpointing alarms as they occur within a building and guiding quick and appropriate response to that location. In addition, management of fan systems to control smoke in a building during a fire is facilitated with a system that combines energy management and fire monitoring functions.

Note, however, that the incorporation of fire, security and energy management functions into a single system increases the complexity of that system. This can result in longer start-up time for the initial installation and more complicated troubleshooting if problems occur. Since the function of fire monitoring is critical to building operation, these disadvantages must be weighed against the previously mentioned advantages to determine if a combined BAS is desired.

DATA TRANSMISSION METHODS

A number of different transmission systems can be used in an EMS for communications between the CPU and FID panels. These transmission systems include telephone lines, coaxial cables, electrical power lines, radio frequency, fiber optics and microwave. Table 11-1 compares the various transmission methods.

Twisted Pair

One of the most common data transmission methods for an EMS is a twisted pair of wires. A twisted pair consists of two insulated conductors twisted together to minimize interference from unwanted signals.

Twisted pairs are permanently hardwired lines between the equipment sending and receiving data that can carry information over a wide range of speeds depending on line characteristics. To maintain a particular data communication rate, the line bandwidth, time delay or the signal to noise ratio may require adjustment by conditioning the line.

Data transmission in twisted pairs, in most cases, is limited to 1200 bps (bits per second) or less. By using signal conditioning, operating speeds up to 9600 bps may be obtained.

Voice Grade Telephone Lines

Voice grade lines used for data transmission are twisted pair circuits defined as type 3002 by the Bell Telephone Company. The local telephone company charges a small connection fee for this service, plus a monthly equipment lease fee. Maintenance is included in the monthly lease fee, with a certain level of service guaranteed.

Two of the major problems involve the quality of telephone pairs provided to the installer and the transmission rate. The minimum quality of each line intended for use in the trunk wiring system must be clearly defined.

The most common voice grade line used for data communication is the unconditioned type 3002 allowing transmission rates up to 1200 bps. The 3002 type line may be used for data transmission up to 9600 bps with the proper line conditioning. Voice grade lines must be used with the same constraints and guidelines for twisted pairs.

Coaxial Cable

Coaxial cable consists of a center conductor surrounded by a shield. The center conductor is separated from the shield by a dielectric. The shield protects against electromagnetic interference. Coaxial cables can operate at data transmission rates in the megabits per second range, but attenuation becomes greater as the data transmission rate increases.

The transmission rate is limited by the data transmission equipment and not by the cable. Regenerative repeaters are required at specific intervals depending on the data rate, nominally every 2000 feet to maintain the signal at usable levels.

Power Line Carrier

Data can be transmitted to remote locations over electric power lines using carrier current transmission that superimposes a low power RF (radio frequency) signal, typically 100 kHz, onto the 60 Hz power distribution system. Since the RF carrier signal cannot operate across transformers, all communicating devices must be connected to the same power circuit (same transformer secondary and phase) unless RF couplers are installed across transformers permitting the transmitters and receivers to be connected over a wider area of the power system. Transmission can be either one-way or two-way.

Note that power line carrier technology is sometimes used in microprocessor based EMS retrofit applications to control single loads in a facility where hardwiring would be difficult and expensive (eg. wiring between two buildings). Figure 11-10 shows a basic power line carrier system configuration.

Radio Frequency

Modulated RF signals, usually VHF or FM radio, can be used as a data transmission method with the installation of radio receivers and transmitters. RF systems can be effectively used for two-way communication between CPU and FID panels where other data transmission methods are not available or suitable for the application. One-way RF systems can be effectively used to control loads at remote locations such as warehouses, unitary heaters and for family housing projects.

The use of RF at a facility however, must be considered carefully to avoid conflict with other existing or planned facility RF systems. Additionally there may be a difficulty in finding a frequency on which to transmit, since there are a limited number available.

The kinds of signals sent over an FM radio system also are limited, as are the distances over which the signals can be transmitted. The greater the distance, the greater the likelihood that erroneous signals will be received.

Fiber Optics

Fiber optics uses the wideband properties of infrared light traveling through transparent fibers. Fiber optics is a reliable communications media which is rapidly becoming cost competitive when compared to other high speed data transmission methods.

The bandwidth of this media is virtually unlimited and extremely high transmission rates can be obtained. The signal attenuation of high quality fiber optic cable is far lower than the best coaxial cables. Repeaters required nominally every 2000 feet for coaxial cable, are 3 to 6 miles apart in fiber optics systems.

Fiber optics terminal equipment selection is limited to date, and there is a lack of skilled installers and maintenance personnel familiar with this media. Fiber optics must be carefully installed and cannot be bent at right angles.

Microware Transmission

For long distance transmission, a microwave link can be used. The primary drawback of microwave links is first cost. Receivers/transmitters are needed at each building in a multi-facility arrangement.

Microwave transmission rates are very fast and are compatible with present and future data requirements. Reliability is excellent, too, but knowledgable maintenance personnel are required. The only limit on expansion is cost.

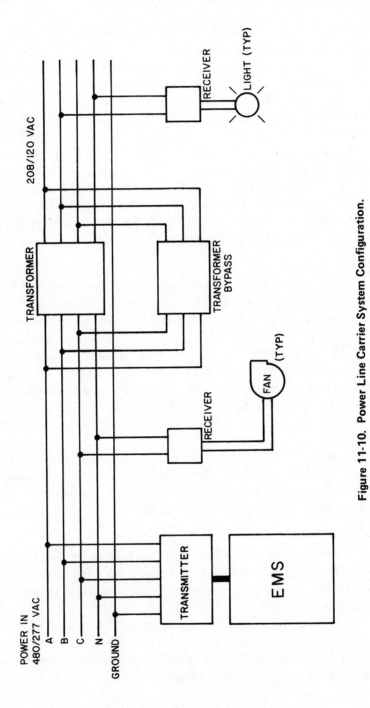

Figure 11-10. Power Line Carrier System Configuration.

Table 11-1. Transmission Method Comparisons

Method	First Cost	Scan Rates	Reliability	Maint. Effort	Expandibility	Compatibility with Future Requirements
Coaxial	high	fast	excellent	min.	unlimited	unlimited
Twisted pair	low	med.	very good	min.	unlimited	limited
RF	med.	fast but limited	low	high	very limited	very limited
Microwave	very high	very fast	excellent	high	unlimited	unlimited
Telephone	very low	slow	low to high	min.	limited	limited
Fiber optics	high	very fast	excellent	min.	unlimited	unlimited
Power Line Carrier	med.	med.	med.	high	limited	limited

SUMMARY

The term energy management system denotes equipment whose functions can range from simple timeclock control to sophisticated building automation. Two broad and overlapping categories of these systems are microprocessor and mini-computer based.

Capabilities of EMS can include scheduling, duty cycling, demand limiting, optimal start, monitoring, direct digital control, fire detection and security.

Direct digital control capability enables the EMS to replace the environmental control system so that it directly manages HVAC operations.

REFERENCES

1. McPartland J.F., Handbook of Practical Electrical Design, McGraw-Hill Book Company, 1984.

2. Smith R.J., Circuits Devices and Systems, John Wiley & Sons, Inc., 1977

3. Industrial Lighting Handbook, National Lighting Bureau, Washington, D.C.

4. Traister, J.E., Practical Lighting Applications for Building Construction, Van Nostrand Reinhold Co., N.Y., N.Y., 1982

5. Sorcar, P.C., Energy Saving Lighting Systems, Van Nostrand Reinhold Co., N.Y., N.Y., 1982

6. Illuminating Engineering Society Lighting Handbook, Illuminating Engineering Society, N.Y., N.Y., 1981

7. Getting the Most From Your Lighting Dollar, National Lighting Bureau, Washington, D.C.

8. Energy Monitoring and Control Systems (EMCS), ARMY TM-815-2, Departments of the Army and Air Force, June, 1983

9. Ottavianio V.B., Energy Management, Ottaviano Technical Services Inc., 1983

10. National Electrical Code, National Fire Protection Association, Boston, Mass., 1984

11. Installation and Owner's Manual, Trimax Controls Inc., Sunnyvale, CA, 1983

12. Electrical Energy Controls, National Electrical Contractors Association, Inc., 1978

13. Thumann, A., Fundamentals of Energy Engineering, Fairmont Press, 1984.

14. Thumann, A., Plant Engineers & Managers Guide to Energy Conservation, 2nd Edition, 1983, Van Nostrand Reinhold Co., N.Y., N.Y.

15. The Engineering Basics of Power Factor Improvement, *Specifying Engineer*, February, 1975; May, 1975.

16. A New Look at Load Shedding, A. Thumann, *Electrical Consultant*, August, 1974.

17. Improving Plant Power Factor, A. Thumann, *Electrical Consultant*, July, 1974.

18. An Efficient Selection of Modern Energy—Saving Light Sources Can Mean Saving of 10% to 30% Power Consumption, H.A. Anderson, *Electrical Consultant*, April, 1974.

19. Electric Power Distribution for Industrial Plants, The Institute of Electrical Electronic Engineers.

APPENDIX I

Table 1. 10% Interest Factor

Period n	Single-payment compound-amount (SPCA)	Single-payment present-worth (SPPW)	Uniform-series compound-amount (USCA)	Sinking-fund payment (SFP)	Capital recovery (CR)	Uniform-series present-worth (USPW)
	Future value of $1 $(1 + i)^n$	Present value of $1 $\dfrac{1}{(1 + i)^n}$	Future value of uniform series of $1 $\dfrac{(1 + i)^n - 1}{i}$	Uniform series whose future value is $1 $\dfrac{i}{(1 + i)^n - 1}$	Uniform series with present value of $1 $\dfrac{i(1 + i)^n}{(1 + i)^n - 1}$	Present value of uniform series of $1 $\dfrac{(1 + i)^n - 1}{i(1 + i)^n}$
1	1.100	0.9091	1.000	1.00000	1.10000	0.909
2	1.210	0.8264	2.100	0.47619	0.57619	1.736
3	1.331	0.7513	3.310	0.30211	0.40211	2.487
4	1.464	0.6830	4.641	0.21547	0.31547	3.170
5	1.611	0.6209	6.105	0.16380	0.26380	3.791
6	1.772	0.5645	7.716	0.12961	0.22961	4.355
7	1.949	0.5132	9.487	0.10541	0.20541	4.868
8	2.144	0.4665	11.436	0.08744	0.18744	5.335
9	2.358	0.4241	13.579	0.07364	0.17364	5.759
10	2.594	0.3855	15.937	0.06275	0.16275	6.144
11	2.853	0.3505	18.531	0.05396	0.15396	6.495
12	3.138	0.3186	21.384	0.04676	0.14676	6.814
13	3.452	0.2897	24.523	0.04078	0.14078	7.103
14	3.797	0.2633	27.975	0.03575	0.13575	7.367
15	4.177	0.2394	31.772	0.03147	0.13147	7.606
16	4.595	0.2176	35.950	0.02782	0.12782	7.824
17	5.054	0.1978	40.545	0.02466	0.12466	8.022
18	5.560	0.1799	45.599	0.02193	0.12193	8.201
19	6.116	0.1635	51.159	0.01955	0.11955	8.365
20	6.727	0.1486	57.275	0.01746	0.11746	8.514
21	7.400	0.1351	64.002	0.01562	0.11562	8.649
22	8.140	0.1228	71.403	0.01401	0.11401	8.772
23	8.954	0.1117	79.543	0.01257	0.11257	8.883
24	9.850	0.1015	88.497	0.01130	0.11130	8.985
25	10.835	0.0923	98.347	0.01017	0.11017	9.077
26	11.918	0.0839	109.182	0.00916	0.10916	9.161
27	13.110	0.0763	121.100	0.00826	0.10826	9.237
28	14.421	0.0693	134.210	0.00745	0.10745	9.307
29	15.863	0.0630	148.631	0.00673	0.10673	9.370
30	17.449	0.0573	164.494	0.00608	0.10608	9.427
35	28.102	0.0356	271.024	0.00369	0.10369	9.644
40	45.259	0.0221	442.593	0.00226	0.10226	9.779
45	72.890	0.0137	718.905	0.00139	0.10139	9.863
50	117.391	0.0085	1163.909	0.00086	0.10086	9.915
55	189.059	0.0053	1880.591	0.00053	0.10053	9.947
60	304.482	0.0033	3034.816	0.00033	0.10033	9.967
65	490.371	0.0020	4893.707	0.00020	0.10020	9.980
70	789.747	0.0013	7887.470	0.00013	0.10013	9.987
75	1271.895	0.0008	12708.954	0.00008	0.10008	9.992
80	2048.400	0.0005	20474.002	0.00005	0.10005	9.995
85	3298.969	0.0003	32979.690	0.00003	0.10003	9.997
90	5313.023	0.0002	53120.226	0.00002	0.10002	9.998
95	8556.676	0.0001	85556.760	0.00001	0.10001	9.999

Table 2. 15% Interest Factor

Period n	Single-payment compound-amount (SPCA) — Future value of $1 $(1 + i)^n$	Single-payment present-worth (SPPW) — Present value of 1 $\dfrac{1}{(1 + i)^n}$	Uniform-series compound-amount (USCA) — Future value of uniform series of $1 $\dfrac{(1 + i)^n - 1}{i}$	Sinking-fund payment (SFP) — Uniform series whose future value is $1 $\dfrac{i}{(1 + i)^n - 1}$	Capital recovery (CR) — Uniform series with present value of $1 $\dfrac{i(1 + i)^n}{(1 + i)^n - 1}$	Uniform-series present-worth (USPW) — Present value of uniform series of $1 $\dfrac{(1 + i)^n - 1}{i(1 + i)^n}$
1	1.120	0.8929	1.000	1.00000	1.12000	0.893
2	1.254	0.7972	2.120	0.47170	0.59170	1.690
3	1.405	0.7118	3.374	0.29635	0.41635	2.402
4	1.574	0.6355	4.779	0.20923	0.32923	3.037
5	1.762	0.5674	6.353	0.15741	0.27741	3.605
6	1.974	0.5066	8.115	0.12323	0.24323	4.111
7	2.211	0.4523	10.089	0.09912	0.21912	4.564
8	2.476	0.4039	12.300	0.08130	0.20130	4.968
9	2.773	0.3606	14.776	0.06768	0.18768	5.328
10	3.106	0.3220	17.549	0.05698	0.17698	5.650
11	3.479	0.2875	20.655	0.04842	0.16842	5.938
12	3.896	0.2567	24.133	0.04144	0.16144	6.194
13	4.363	0.2292	28.029	0.03568	0.15568	6.424
14	4.887	0.2046	32.393	0.03087	0.15087	6.628
15	5.474	0.1827	37.280	0.02682	0.14682	6.811
16	6.130	0.1631	42.753	0.02339	0.14339	6.974
17	6.866	0.1456	48.884	0.02046	0.14046	7.120
18	7.690	0.1300	55.750	0.01794	0.13794	7.250
19	8.613	0.1161	63.440	0.01576	0.13576	7.366
20	9.646	0.1037	72.052	0.01388	0.13388	7.469
21	10.804	0.0926	81.699	0.01224	0.13224	7.562
22	12.100	0.0826	92.503	0.01081	0.13081	7.645
23	13.552	0.0738	104.603	0.00956	0.12956	7.718
24	15.179	0.0659	118.155	0.00846	0.12846	7.784
25	17.000	0.0588	133.334	0.00750	0.12750	7.843
26	19.040	0.0525	150.334	0.00665	0.12665	7.896
27	21.325	0.0469	169.374	0.00590	0.12590	7.943
28	23.884	0.0419	190.699	0.00524	0.12524	7.984
29	26.750	0.0374	214.583	0.00466	0.12466	8.022
30	29.960	0.0334	241.333	0.00414	0.12414	8.055
35	52.800	0.0189	431.663	0.00232	0.12232	8.176
40	93.051	0.0107	767.091	0.00130	0.12130	8.244
45	163.988	0.0061	1358.230	0.00074	0.12074	8.283
50	289.002	0.0035	2400.018	0.00042	0.12042	8.304
55	509.321	0.0020	4236.005	0.00024	0.12024	8.317
60	897.597	0.0011	7471.641	0.00013	0.12013	8.324
65	1581.872	0.0006	13173.937	0.00008	0.12008	8.328
70	2787.800	0.0004	23223.332	0.00004	0.12004	8.330
75	4913.056	0.0002	40933.799	0.00002	0.12002	8.332
80	8658.483	0.0001	72145.692	0.00001	0.12001	8.332

Table 3. 20% Interest Factor

Period n	Single-payment compound-amount (SPCA) Future value of $1 $(1+i)^n$	Single-payment present-worth (SPPW) Present value of 1 $\dfrac{1}{(1+i)^n}$	Uniform-series compound-amount (USCA) Future value of uniform series of $1 $\dfrac{(1+i)^n-1}{i}$	Sinking-fund payment (SFP) Uniform series whose future value is $1 $\dfrac{i}{(1+i)^n-1}$	Capital recovery (CR) Uniform series with present value of $1 $\dfrac{i(1+i)^n}{(1+i)^n-1}$	Uniform-series present-worth (USPW) Present value of uniform series of $1 $\dfrac{(1+i)^n-1}{i(1+i)^n}$
1	1.150	0.8696	1.000	1.00000	1.15000	0.870
2	1.322	0.7561	2.150	0.46512	0.61512	1.626
3	1.521	0.6575	3.472	0.28798	0.43798	2.283
4	1.749	0.5718	4.993	0.20027	0.35027	2.855
5	2.011	0.4972	6.742	0.14832	0.29832	3.352
6	2.313	0.4323	8.754	0.11424	0.26424	3.784
7	2.660	0.3759	11.067	0.09036	0.24036	4.160
8	3.059	0.3269	13.727	0.07285	0.22285	4.487
9	3.518	0.2843	16.786	0.05957	0.20957	4.772
10	4.046	0.2472	20.304	0.04925	0.19925	5.019
11	4.652	0.2149	24.349	0.04107	0.19107	5.234
12	5.350	0.1869	29.002	0.03448	0.18448	5.421
13	6.153	0.1625	34.352	0.02911	0.17911	5.583
14	7.076	0.1413	40.505	0.02469	0.17469	5.724
15	8.137	0.1229	47.580	0.02102	0.17102	5.847
16	9.358	0.1069	55.717	0.01795	0.16795	5.954
17	10.761	0.0929	65.075	0.01537	0.16537	6.047
18	12.375	0.0808	75.836	0.01319	0.16319	6.128
19	14.232	0.0703	88.212	0.01134	0.16134	6.198
20	16.367	0.0611	102.444	0.00976	0.15976	6.259
21	18.822	0.0531	118.810	0.00842	0.15842	6.312
22	21.645	0.0462	137.632	0.00727	0.15727	6.359
23	24.891	0.0402	159.276	0.00628	0.15628	6.399
24	28.625	0.0349	184.168	0.00543	0.15543	6.434
25	32.919	0.0304	212.793	0.00470	0.15470	6.464
26	37.857	0.0264	245.712	0.00407	0.15407	6.491
27	43.535	0.0230	283.569	0.00353	0.15353	6.514
28	50.066	0.0200	327.104	0.00306	0.15306	6.534
29	57.575	0.0174	377.170	0.00265	0.15265	6.551
30	66.212	0.0151	434.745	0.00230	0.15230	6.566
35	133.176	0.0075	881.170	0.00113	0.15113	6.617
40	267.864	0.0037	1779.090	0.00056	0.15056	6.642
45	538.769	0.0019	3585.128	0.00028	0.15028	6.654
50	1083.657	0.0009	7217.716	0.00014	0.15014	6.661
55	2179.622	0.0005	14524.148	0.00007	0.15007	6.664
60	4383.999	0.0002	29219.992	0.00003	0.15003	6.665
65	8817.787	0.0001	58778.583	0.00002	0.15002	6.666

Table 4. 20% Interest Factor

Period n	Single-payment compound-amount (SPCA) Future value of $1 $(1 + i)^n$	Single-payment present-worth (SPPW) Present value of $1 $\dfrac{1}{(1 + i)^n}$	Uniform-series compound-amount (USCA) Future value of uniform series of $1 $\dfrac{(1 + i)^n - 1}{i}$	Sinking-fund payment (SFP) Uniform series whose future value is $1 $\dfrac{i}{(1 + i)^n - 1}$	Capital recovery (CR) Uniform series with present value of $1 $\dfrac{i(1 + i)^n}{(1 + i)^n - 1}$	Uniform-series present-worth (USPW) Present value of uniform series of $1 $\dfrac{(1 + i)^n - 1}{i(1 + i)^n}$
1	1.200	0.8333	1.000	1.00000	1.20000	0.833
2	1.440	0.6944	2.200	0.45455	0.65455	1.528
3	1.728	0.5787	3.640	0.27473	0.47473	2.106
4	2.074	0.4823	5.368	0.18629	0.38629	2.589
5	2.488	0.4019	7.442	0.13438	0.33438	2.991
6	2.986	0.3349	9.930	0.10071	0.30071	3.326
7	3.583	0.2791	12.916	0.07742	0.27742	3.605
8	4.300	0.2326	16.499	0.06061	0.26061	3.837
9	5.160	0.1938	20.799	0.04808	0.24808	4.031
10	6.192	0.1615	25.959	0.03852	0.23852	4.192
11	7.430	0.1346	32.150	0.03110	0.23110	4.327
12	8.916	0.1122	39.581	0.02526	0.22526	4.439
13	10.699	0.0935	48.497	0.02062	0.22062	4.533
14	12.839	0.0779	59.196	0.01689	0.21689	4.611
15	15.407	0.0649	72.035	0.01388	0.21388	4.675
16	18.488	0.0541	87.442	0.01144	0.21144	4.730
17	22.186	0.0451	105.931	0.00944	0.20944	4.775
18	26.623	0.0376	128.117	0.00781	0.20781	4.812
19	31.948	0.0313	154.740	0.00646	0.20646	4.843
20	38.338	0.0261	186.688	0.00536	0.20536	4.870
21	46.005	0.0217	225.026	0.00444	0.20444	4.891
22	55.206	0.0181	271.031	0.00369	0.20369	4.909
23	66.247	0.0151	326.237	0.00307	0.20307	4.925
24	79.497	0.0126	392.484	0.00255	0.20255	4.937
25	95.396	0.0105	471.981	0.00212	0.20212	4.948
26	114.475	0.0087	567.377	0.00176	0.20176	4.956
27	137.371	0.0073	681.853	0.00147	0.20147	4.964
28	164.845	0.0061	819.223	0.00122	0.20122	4.970
29	197.814	0.0051	984.068	0.00102	0.20102	4.975
30	237.376	0.0042	1181.882	0.00085	0.20085	4.979
35	590.668	0.0017	2948.341	0.00034	0.20034	4.992
40	1469.772	0.0007	7343.858	0.00014	0.20014	4.997
45	3657.262	0.0003	18281.310	0.00005	0.20005	4.999
50	9100.438	0.0001	45497.191	0.00002	0.20002	4.999

Table 5. 25% Interest Factor

Period n	Single-payment compound-amount (SPCA) Future value of \$1 $(1 + i)^n$	Single-payment present-worth (SPPW) Present value of \$1 $\dfrac{1}{(1 + i)^n}$	Uniform-series compound amount (USCA) Future value of uniform series of \$1 $\dfrac{(1 + i)^n - 1}{i}$	Sinking-fund payment (SFP) Uniform series whose future value is \$1 $\dfrac{i}{(1 + i)^n - 1}$	Capital recovery (CR) Uniform series with present value of \$1 $\dfrac{i(1 + i)^n}{(1 + i)^n - 1}$	Uniform-series present-worth (USPW) Present value of uniform series of \$1 $\dfrac{(1 + i)^n - 1}{i(1 + i)^n}$
1	1.250	0.8000	1.000	1.00000	1.25000	0.800
2	1.562	0.6400	2.250	0.44444	0.69444	1.440
3	1.953	0.5120	3.812	0.26230	0.51230	1.952
4	2.441	0.4096	5.766	0.17344	0.42344	2.362
5	3.052	0.3277	8.207	0.12185	0.37185	2.689
6	3.815	0.2621	11.259	0.08882	0.33882	2.951
7	4.768	0.2097	15.073	0.06634	0.31634	3.161
8	5.960	0.1678	19.842	0.05040	0.30040	3.329
9	7.451	0.1342	25.802	0.03876	0.28876	3.463
10	9.313	0.1074	33.253	0.03007	0.28007	3.571
11	11.642	0.0859	42.566	0.02349	0.27349	3.656
12	14.552	0.0687	54.208	0.01845	0.26845	3.725
13	18.190	0.0550	68.760	0.01454	0.26454	3.780
14	22.737	0.0440	86.949	0.01150	0.26150	3.824
15	28.422	0.0352	109.687	0.00912	0.25912	3.859
16	35.527	0.0281	138.109	0.00724	0.25724	3.887
17	44.409	0.0225	173.636	0.00576	0.25576	3.910
18	55.511	0.0180	218.045	0.00459	0.25459	3.928
19	69.389	0.0144	273.556	0.00366	0.25366	3.942
20	86.736	0.0115	342.945	0.00292	0.25292	3.954
21	108.420	0.0092	429.681	0.00233	0.25233	3.963
22	135.525	0.0074	538.101	0.00186	0.25186	3.970
23	169.407	0.0059	673.626	0.00148	0.25148	3.976
24	211.758	0.0047	843.033	0.00119	0.25119	3.981
25	264.698	0.0038	1054.791	0.00095	0.25095	3.985
26	330.872	0.0030	1319.489	0.00076	0.25076	3.988
27	413.590	0.0024	1650.361	0.00061	0.25061	3.990
28	516.988	0.0019	2063.952	0.00048	0.25048	3.992
29	646.235	0.0015	2580.939	0.00039	0.25039	3.994
30	807.794	0.0012	3227.174	0.00031	0.25031	3.995
35	2465.190	0.0004	9856.761	0.00010	0.25010	3.998
40	7523.164	0.0001	30088.655	0.00003	0.25003	3.999

Table 6. 30% Interest Factor

Period n	Single-payment compound-amount (SPCA) Future value of \$1 $(1 + i)^n$	Single-payment present-worth (SPPW) Present value of \$1 $\dfrac{1}{(1 + i)^n}$	Uniform-series compound-amount (USCA) Future value of uniform series of \$1 $\dfrac{(1 + i)^n - 1}{i}$	Sinking-fund payment (SFP) Uniform series whose future value is \$1 $\dfrac{i}{(1 + i)^n - 1}$	Capital recovery (CR) Uniform series with present value of \$1 $\dfrac{i(1 + i)^n}{(1 + i)^n - 1}$	Uniform-series present-worth (USPW) Present value of uniform series of \$1 $\dfrac{(1 + i)^n - 1}{i(1 + i)^n}$
1	1.300	0.7692	1.000	1.00000	1.30000	0.769
2	1.690	0.5917	2.300	0.43478	0.73478	1.361
3	2.197	0.4552	3.990	0.25063	0.55063	1.816
4	2.856	0.3501	6.187	0.16163	0.46163	2.166
5	3.713	0.2693	9.043	0.11058	0.41058	2.436
6	4.827	0.2072	12.756	0.07839	0.37839	2.643
7	6.275	0.1594	17.583	0.05687	0.35687	2.802
8	8.157	0.1226	23.858	0.04192	0.34192	2.925
9	10.604	0.0943	32.015	0.03124	0.33124	3.019
10	13.786	0.0725	42.619	0.02346	0.32346	3.092
11	17.922	0.0558	56.405	0.01773	0.31773	3.147
12	23.298	0.0429	74.327	0.01345	0.31345	3.190
13	30.288	0.0330	97.625	0.01024	0.31024	3.223
14	39.374	0.0254	127.913	0.00782	0.30782	3.249
15	51.186	0.0195	167.286	0.00598	0.30598	3.268
16	66.542	0.0150	218.472	0.00458	0.30458	3.283
17	86.504	0.0116	285.014	0.00351	0.30351	3.295
18	112.455	0.0089	371.518	0.00269	0.30269	3.304
19	146.192	0.0068	483.973	0.00207	0.30207	3.311
20	190.050	0.0053	630.165	0.00159	0.30159	3.316
21	247.065	0.0040	820.215	0.00122	0.30122	3.320
22	321.184	0.0031	1067.280	0.00094	0.30094	3.323
23	417.539	0.0024	1388.464	0.00072	0.30072	3.325
24	542.801	0.0018	1806.003	0.00055	0.30055	3.327
25	705.641	0.0014	2348.803	0.00043	0.30043	3.329
26	917.333	0.0011	3054.444	0.00033	0.30033	3.330
27	1192.533	0.0008	3971.778	0.00025	0.30025	3.331
28	1550.293	0.0006	5164.311	0.00019	0.30019	3.331
29	2015.381	0.0005	6714.604	0.00015	0.30015	3.332
30	2619.996	0.0004	8729.985	0.00011	0.30011	3.332
35	9727.860	0.0001	32422.868	0.00003	0.30003	3.333

Table 7. 40% Interest Factor

Period n	Single-payment compound-amount (SPCA) Future value of $1 $(1 + i)^n$	Single-payment present-worth (SPPW) Present value of $1 $\dfrac{1}{(1 + i)^n}$	Uniform-series compound-amount (USCA) Future value of uniform series of $1 $\dfrac{(1 + i)^n - 1}{i}$	Sinking-fund payment (SFP) Uniform series whose future value is $1 $\dfrac{i}{(1 + i)^n - 1}$	Capital recovery (CR) Uniform series with present value of $1 $\dfrac{i(1 + i)^n}{(1 + i)^n - 1}$	Uniform-series present-worth (USPW) Present value of uniform series of $1 $\dfrac{(1 + i)^n - 1}{i(1 + i)^n}$
1	1.400	0.7143	1.000	1.00000	1.40000	0.714
2	1.960	0.5102	2.400	0.41667	0.81667	1.224
3	2.744	0.3644	4.360	0.22936	0.62936	1.589
4	3.842	0.2603	7.104	0.14077	0.54077	1.849
5	5.378	0.1859	10.946	0.09136	0.49136	2.035
6	7.530	0.1328	16.324	0.06126	0.46126	2.168
7	10.541	0.0949	23.853	0.04192	0.44192	2.263
8	14.758	0.0678	34.395	0.02907	0.42907	2.331
9	20.661	0.0484	49.153	0.02034	0.42034	2.379
10	28.925	0.0346	69.814	0.01432	0.41432	2.414
11	40.496	0.0247	98.739	0.01013	0.41013	2.438
12	56.694	0.0176	139.235	0.00718	0.40718	2.456
13	79.371	0.0126	195.929	0.00510	0.40510	2.469
14	111.120	0.0090	275.300	0.00363	0.40363	2.478
15	155.568	0.0064	386.420	0.00259	0.40259	2.484
16	217.795	0.0046	541.988	0.00185	0.40185	2.489
17	304.913	0.0033	759.784	0.00132	0.40132	2.492
18	426.879	0.0023	1064.697	0.00094	0.40094	2.494
19	597.630	0.0017	1491.576	0.00067	0.40067	2.496
20	836.683	0.0012	2089.206	0.00048	0.40048	2.497
21	1171.356	0.0009	2925.889	0.00034	0.40034	2.498
22	1639.898	0.0006	4097.245	0.00024	0.40024	2.498
23	2295.857	0.0004	5737.142	0.00017	0.40017	2.499
24	3214.200	0.0003	8032.999	0.00012	0.40012	2.499
25	4499.880	0.0002	11247.199	0.00009	0.40009	2.499
26	6299.831	0.0002	15747.079	0.00006	0.40006	2.500
27	8819.764	0.0001	22046.910	0.00005	0.40005	2.500

Table 8. 50% Interest Factor

Period n	Single-payment compound-amount (SPCA) Future value of $1 $(1 + i)^n$	Single-payment present-worth (SPPW) Present value of $1 $\dfrac{1}{(1 + i)^n}$	Uniform-series compound-amount (USCA) Future value of uniform series of $1 $\dfrac{(1 + i)^n - 1}{i}$	Sinking-fund payment (SFP) Uniform series whose future value is $1 $\dfrac{i}{(1 + i)^n - 1}$	Capital recovery (CR) Uniform series with present value of $1 $\dfrac{i(1 + i)^n}{(1 + i)^n - 1}$	Uniform-series present-worth (USPW) Present value of uniform series of $1 $\dfrac{(1 + i)^n - 1}{i(1 + i)^n}$
1	1.500	0.6667	1.000	1.00000	1.50000	0.667
2	2.250	0.4444	2.500	0.40000	0.90000	1.111
3	3.375	0.2963	4.750	0.21053	0.71053	1.407
4	5.062	0.1975	8.125	0.12308	0.62308	1.605
5	7.594	0.1317	13.188	0.07583	0.57583	1.737
6	11.391	0.0878	20.781	0.04812	0.54812	1.824
7	17.086	0.0585	32.172	0.03108	0.53108	1.883
8	25.629	0.0390	49.258	0.02030	0.52030	1.922
9	38.443	0.0260	74.887	0.01335	0.51335	1.948
10	57.665	0.0173	113.330	0.00882	0.50882	1.965
11	86.498	0.0116	170.995	0.00585	0.50585	1.977
12	129.746	0.0077	257.493	0.00388	0.50388	1.985
13	194.620	0.0051	387.239	0.00258	0.50258	1.990
14	291.929	0.0034	581.859	0.00172	0.50172	1.993
15	437.894	0.0023	873.788	0.00114	0.50114	1.995
16	656.841	0.0015	1311.682	0.00076	0.50076	1.997
17	985.261	0.0010	1968.523	0.00051	0.50051	1.998
18	1477.892	0.0007	2953.784	0.00034	0.50034	1.999
19	2216.838	0.0005	4431.676	0.00023	0.50023	1.999
20	3325.257	0.0003	6648.513	0.00015	0.50015	1.999
21	4987.885	0.0002	9973.770	0.00010	0.50010	2.000
22	7481.828	0.0001	14961.655	0.00007	0.50007	2.000

Table 9. Five-Year Escalation Table

Present Worth of a Series of Escalating Payments Compounded Annually
Discount-Escalation Factors for n = 5 Years

Discount Rate	Annual Escalation Rate					
	0.10	0.12	0.14	0.16	0.18	0.20
0.10	5.000000	5.279234	5.572605	5.880105	6.202627	6.540569
0.11	4.866862	5.136200	5.420152	5.717603	6.029313	6.355882
0.12	4.738562	5.000000	5.274242	5.561868	5.863289	6.179066
0.13	4.615647	4.869164	5.133876	5.412404	5.704137	6.009541
0.14	4.497670	4.742953	5.000000	5.269208	5.551563	5.847029
0.15	4.384494	4.622149	4.871228	5.131703	5.404955	5.691165
0.16	4.275647	4.505953	4.747390	5.000000	5.264441	5.541511
0.17	4.171042	4.394428	4.628438	4.873699	5.129353	5.397964
0.18	4.070432	4.287089	4.513947	4.751566	5.000000	5.259749
0.19	3.973684	4.183921	4.403996	4.634350	4.875619	5.126925
0.20	3.880510	4.084577	4.298207	4.521178	4.755725	5.000000
0.21	3.790801	3.989001	4.196400	4.413341	4.640260	4.877689
0.22	4.704368	3.896891	4.098287	4.308947	4.529298	4.759649
0.23	3.621094	3.808179	4.003835	4.208479	4.422339	4.645864
0.24	3.540773	3.722628	3.912807	4.111612	4.319417	4.536517
0.25	3.463301	3.640161	3.825008	4.018249	4.220158	4.431144
0.26	3.388553	3.560586	3.740376	3.928286	4.124553	4.329514
0.27	3.316408	3.483803	3.658706	3.841442	4.032275	4.231583
0.28	3.246718	3.409649	3.579870	3.757639	3.943295	4.137057
0.29	3.179393	3.338051	3.503722	3.676771	3.857370	4.045902
0.30	3.114338	3.268861	3.430201	3.598653	3.774459	3.957921
0.31	3.051452	3.201978	3.359143	3.523171	3.694328	3.872901
0.32	2.990618	3.137327	3.290436	3.450224	3.616936	3.790808
0.33	2.931764	3.074780	3.224015	3.379722	3.542100	3.711472
0.34	2.874812	3.014281	3.159770	3.311524	3.469775	3.634758

Table 10. Ten-Year Escalation Table

Present Worth of a Series of Escalating Payments Compounded Annually
Discount-Escalation Factors for $n = 10$ Years

Discount Rate	Annual Escalation Rate					
	0.10	0.12	0.14	0.16	0.18	0.20
0.10	10.000000	11.056250	12.234870	13.548650	15.013550	16.646080
0.11	9.518405	10.508020	11.613440	12.844310	14.215140	15.741560
0.12	9.068870	10.000000	11.036530	12.190470	13.474590	14.903510
0.13	8.650280	9.526666	10.498990	11.582430	12.786980	14.125780
0.14	8.259741	9.084209	10.000000	11.017130	12.147890	13.403480
0.15	7.895187	8.672058	9.534301	10.490510	11.552670	12.731900
0.16	7.554141	8.286779	9.099380	10.000000	10.998720	12.106600
0.17	7.234974	7.926784	8.693151	9.542653	10.481740	11.524400
0.18	6.935890	7.589595	8.312960	9.113885	10.000000	10.980620
0.19	6.655455	7.273785	7.957330	8.713262	9.549790	10.472990
0.20	6.392080	6.977461	7.624072	8.338518	9.128122	10.000000
0.21	6.144593	6.699373	7.311519	7.987156	8.733109	9.557141
0.22	5.911755	6.437922	7.017915	7.657542	8.363208	9.141752
0.23	5.692557	6.192047	6.742093	7.348193	8.015993	8.752133
0.24	5.485921	5.960481	6.482632	7.057347	7.690163	8.387045
0.25	5.290990	5.742294	6.238276	6.783767	7.383800	8.044173
0.26	5.106956	5.536463	6.008083	6.526298	7.095769	7.721807
0.27	4.933045	5.342146	5.790929	6.283557	6.824442	7.418647
0.28	4.768518	5.158489	5.585917	6.054608	6.568835	7.133100
0.29	4.612762	4.984826	5.392166	5.838531	6.327682	6.864109
0.30	4.465205	4.820429	5.209000	5.634354	6.100129	6.610435
0.31	4.325286	4.664669	5.035615	5.441257	5.885058	6.370867
0.32	4.192478	4.517015	4.871346	5.258512	5.681746	6.144601
0.33	4.066339	4.376884	4.715648	5.085461	5.489304	5.930659
0.34	3.946452	4.243845	4.567942	4.921409	5.307107	5.728189

Table 11. Fifteen-Year Escalation Table

Present Worth of a Series of Escalating Payments Compounded Annually
Discount-Escalation Factors for *n* = 15 years

Discount Rate	Annual Escalation Rate					
	0.10	0.12	0.14	0.16	0.18	0.20
0.10	15.000000	17.377880	20.199780	23.549540	27.529640	32.259620
0.11	13.964150	16.126230	18.690120	21.727370	25.328490	29.601330
0.12	13.026090	15.000000	17.332040	20.090360	23.355070	27.221890
0.13	12.177030	13.981710	16.105770	18.616160	21.581750	25.087260
0.14	11.406510	13.057790	15.000000	17.287320	19.985530	23.169060
0.15	10.706220	12.220570	13.998120	16.086500	18.545150	21.442230
0.16	10.068030	11.459170	13.088900	15.000000	17.244580	19.884420
0.17	9.485654	10.766180	12.262790	14.015480	16.066830	18.477610
0.18	8.953083	10.133630	11.510270	13.118840	15.000000	17.203010
0.19	8.465335	9.555676	10.824310	12.303300	14.030830	16.047480
0.20	8.017635	9.026333	10.197550	11.560150	13.148090	15.000000
0.21	7.606115	8.540965	9.623969	10.881130	12.343120	14.046400
0.22	7.227109	8.094845	9.097863	10.259820	11.608480	13.176250
0.23	6.877548	7.684317	8.614813	9.690559	10.936240	12.381480
0.24	6.554501	7.305762	8.170423	9.167798	10.320590	11.655310
0.25	6.255518	6.956243	7.760848	8.687104	9.755424	10.990130
0.26	5.978393	6.632936	7.382943	8.244519	9.236152	10.379760
0.27	5.721101	6.333429	7.033547	7.836080	8.757889	9.819020
0.28	5.481814	6.055485	6.710042	7.458700	8.316982	9.302823
0.29	5.258970	5.797236	6.410005	7.109541	7.909701	8.827153
0.30	5.051153	5.556882	6.131433	6.785917	7.533113	8.388091
0.31	4.857052	5.332839	5.872303	6.485500	7.184156	7.982019
0.32	4.675478	5.123753	5.630905	6.206250	6.860492	7.606122
0.33	4.505413	4.928297	5.405771	5.946343	6.559743	7.257569
0.34	4.345926	4.745399	5.195502	5.704048	6.280019	6.933897

Table 12. Twenty-Year Escalation Table

Present Worth of a Series of Escalating Payments Compounded Annually
Discount-Escalation Factors for n = 20 Years

Discount Rate	Annual Escalation Rate					
	0.10	0.12	0.14	0.16	0.18	0.20
0.10	20.000000	24.295450	29.722090	36.592170	45.308970	56.383330
0.11	18.213210	22.002090	26.776150	32.799710	40.417480	50.067940
0.12	16.642370	20.000000	24.210030	29.505400	36.181240	44.614710
0.13	15.259850	18.243100	21.964990	26.634490	32.502270	39.891400
0.14	14.038630	16.694830	20.000000	24.127100	29.298170	35.789680
0.15	12.957040	15.329770	18.271200	21.929940	26.498510	32.218060
0.16	11.995640	14.121040	16.746150	20.000000	24.047720	29.098950
0.17	11.138940	13.048560	15.397670	18.300390	21.894660	26.369210
0.18	10.373120	12.093400	14.201180	16.795710	20.000000	23.970940
0.19	9.686791	11.240870	13.137510	15.463070	18.326720	21.860120
0.20	9.069737	10.477430	12.188860	14.279470	16.844020	20.000000
0.21	8.513605	9.792256	11.340570	13.224610	15.527270	18.353210
0.22	8.010912	9.175267	10.579620	12.282120	14.355520	16.890730
0.23	7.555427	8.618459	9.895583	11.438060	13.309280	15.589300
0.24	7.141531	8.114476	9.278916	10.679810	12.373300	14.429370
0.25	6.764528	7.657278	8.721467	9.997057	11.533310	13.392180
0.26	6.420316	7.241402	8.216490	9.380883	10.778020	12.462340
0.27	6.105252	6.862203	7.757722	8.823063	10.096710	11.626890
0.28	5.816151	6.515563	7.339966	8.316995	9.480940	10.874120
0.29	5.550301	6.198027	6.958601	7.856833	8.922847	10.194520
0.30	5.305312	5.906440	6.609778	7.437339	8.416060	9.579437
0.31	5.079039	5.638064	6.289875	7.054007	7.954518	9.021190
0.32	4.869585	5.390575	5.995840	6.702967	7.533406	8.513612
0.33	4.675331	5.161809	5.725066	6.380829	7.148198	8.050965
0.34	4.494838	4.949990	5.475180	6.084525	6.795200	7.628322

INDEX